ARCHITECTURE IN AN AGE OF UNCERTAINTY

Ashgate Studies in Architecture Series

SERIES EDITOR: EAMONN CANNIFFE, MANCHESTER SCHOOL OF ARCHITECTURE, MANCHESTER METROPOLITAN UNIVERSITY, UK

The discipline of Architecture is undergoing subtle transformation as design awareness permeates our visually dominated culture. Technological change, the search for sustainability and debates around the value of place and meaning of the architectural gesture are aspects which will affect the cities we inhabit. This series seeks to address such topics, both theoretically and in practice, through the publication of high quality original research, written and visual.

Other titles in this series

The Architectural Capriccio
Memory, Fantasy and Invention
Lucien Steil
ISBN 978 1 4094 3191 6

The Architecture of Pleasure
British Amusement Parks 1900–1939
Josephine Kane
ISBN 978 1 4094 1074 4

No Matter: Theories and Practices of the Ephemeral in Architecture
Anastasia Karandinou
ISBN 978 1 4094 6628 4

The Challenge of Emulation in Art and Architecture
Between Imitation and Invention
David Mayernik
ISBN 978 1 4094 5767 1

The Architecture of Edwin Maxwell Fry and Jane Drew
Twentieth Century Architecture, Pioneer Modernism and the Tropics
Iain Jackson and Jessica Holland
ISBN 978 1 4094 5198 3

Forthcoming titles in this series

The Architecture of Industry
Changing Paradigms in Industrial Building and Planning
Mathew Aitchison
ISBN 978 1 4724 3299 5

Architecture in an Age of Uncertainty

Edited by

Benjamin Flowers
Georgia Institute of Technology, USA

Routledge
Taylor & Francis Group

LONDON AND NEW YORK

First published 2014 by Ashgate Publishing

2 Park Square, Milton Park, Abingdon, Oxon, OX14 4RN
605 Third Avenue, New York, NY 10017

Routledge is an imprint of the Taylor & Francis Group, an informa business

First issued in paperback 2020

British Library Cataloguing in Publication Data
A catalogue record for this book is available from the British Library.

Library of Congress Cataloging-in-Publication Data
Architecture in an age of uncertainty : [edited] by Benjamin Flowers.
    pages cm. -- (Ashgate studies in architecture)
    Includes bibliographical references and index.
    ISBN 978-1-4094-4575-3 (hardback) -- ISBN 978-1-4094-4576-0 (ebook) -- ISBN 978-1-4724-0019-2 (epub) 1. Architectural practice. 2. Architecture and society.
    NA1995.A757 2014
                                                                    2014011131

ISBN 13: 978-1-4094-4575-3 (hbk)
ISBN 13: 978-0-367-73962-1 (pbk)

Contents

List of Illustrations

3   Faster Better Cheaper: Social Process for a Modular Future

3.1   The SAGE Classroom, by Portland State University and Blazer Industries

3.2   Typical Modular Classroom in the US

3.3   *Upgrade*, A. Green, C. Bardawil and L. Churchill

3.4   *Dropbox*, J. Tomasini

3.5   The SAGE Classroom, by Portland State University and Blazer Industries

4   Re-defining Architectural Performance—*Survival Through Design* and the Sentient Environmentalism of Richard Neutra

4.1 and 4.2   Exterior and landscape view of the Alfred de Schulthess House, Richard Neutra Architect, built in 1956, in Cubanacan Playa, Havana, Cuba. By permission of the Bundesamt für Bauten und Logistik, Government of Switzerland, Photographer Nathaniel T. Schlundt

5   Lightweight, Impermanent, Recycled

5.1   Placement of pavilions

5.2   The pavilion at night

5.3   Interior view of the pavilion

5.4   Parts of the flexible mold

5.5   Plants inhabit the pavilion's envelope

5.6   Plants inhabit the pavilion's envelope

List of Contributors

JENNIFER BONNER

Jennifer Bonner is an Assistant Professor at Georgia Institute of Technology School of Architecture and Director of Studio Bonner. A graduate of Harvard University Graduate School of Design and Auburn University Rural Studio, her work has been published and exhibited internationally. She is Founder of *A Guide to the Dirty South* with forthcoming titles in Atlanta, New Orleans, and Miami.

CHRISTINE HAVEN CANABOU

Christine Haven Canabou is working on an exhibition that explores disaster-resilient design at the National Building Museum. She is a graduate of the Harvard Graduate School of Design's Master in Architecture program and the Medill School of Journalism at Northwestern University. Previously, she was a staff writer at *Fast Company* magazine.

THOMAS R. FISHER

Thomas R. Fisher is a professor of architecture and the Dean of the College of Design at the University of Minnesota. Educated at Cornell University in architecture and Case Western Reserve University in intellectual history, he previously served as the Regional Preservation Officer at the Western Reserve Historical Society in Cleveland, the Historical Architect of the Connecticut State Historical Commission, and the Editorial Director of Progressive Architecture magazine. He has lectured or juried at over 40 schools and 60 professional societies, and has published 35 book chapters or introductions and over 250 articles.

BENJAMIN FLOWERS

Benjamin Flowers is an Associate Professor in the School of Architecture at the Georgia Institute of Technology. His work examines architecture as a form of social activity situated within the intersecting spheres of politics, culture, and economy. Looking in particular at skyscrapers and stadiums, he focuses on the ways these structures are constructed, the ends to which they are used, and the nature of public reaction to them. His research has attracted recognition and funding from Columbia University's Buell Center for Architecture, Cornell University's John Nolen Fellowship, the Society of Architectural Historians, and the Hagley Museum and Library. His first book, *Skyscraper: The Politics and Power of Building New York City in the Twentieth Century* (University of Pennsylvania Press, 2009), was named a 2010 Outstanding Academic Title in Architecture by *Choice* Magazine. He is currently completing a book on stadia around the world.

GEORGE B. JOHNSTON

George B. Johnston is Professor of Architecture at Georgia Institute of Technology, where he teaches courses in architectural design and in the history and theory of architectural practice. Based upon his background as both an architect and cultural historian, Johnston is especially open to and able to support research and design projects that involve themes of memory and modernity; institutions of cultural exhibition and display; approaches to American vernacular architecture and cultural landscape; and the critique of the everyday. Johnston's research interrogates the social, historical, and cultural implications of making architecture in the American context. His book, *Drafting Culture: A Social History of Architectural Graphic Standards* (MIT Press, 2008), has been lauded for its insights into the ongoing technological transformation of the profession; it received the 2009 Outstanding Book Award from the Southeast Society of Architectural Historians.

MARGARETTE LEITE

Margarette Leite teaches building tectonics, material sustainability and community engaged design at Portland State University's School of Architecture in Portland, Oregon, and is a Fellow of the Center for Public Interest Design there. She is known for her work with local schools and in disaster relief communities. Her work on the SAGE green modular classroom received a 2013 SEED award for social, economic and environmental design.

WILLIAM MANGOLD

William Mangold is an Adjunct Professor in Interior Design at Pratt Institute and a Ph.D. candidate in the Environmental Psychology program at The Graduate Center, CUNY. Trained as an architect, his research looks at social responsibility in design and utopian visions for transforming the social and spatial environment. He has recently completed an edited volume bringing together scholarship from across disciplines: *The People, Place, and Space Reader*.

DAVID MONTEYNE

David Monteyne is an Associate Professor and Associate Dean, Architecture, in the Faculty of Environmental Design at the University of Calgary, Canada. He teaches courses in the history and theory of architecture and urbanism. Monteyne has held fellowships at CRASSH (Centre for Research in the Arts, Social Sciences and Humanities) and Clare Hall, University of Cambridge, and at the Canadian Centre for Architecture, Montreal. He is the author of *Fallout Shelter: Designing for Civil Defense in the Cold War* (Minneapolis: University of Minnesota Press, 2011).

NATHAN RICHARDSON

Assistant Professor Nathan Richardson teaches design, real estate, and entrepreneurship at the Oklahoma State University School of Architecture and holds a joint appointment at the School of Entrepreneurship as a Riata Faculty Fellow. Prior to joining OSU in the fall of 2009, Nathan received his Master of Design Studies with Distinction from the Harvard University Graduate School of Design. Nathan is a licensed architect in Massachusetts and Oklahoma.

GERNOT RIETHER

Gernot Riether is a visiting professor in the Department of Architecture, Ball State University. His work focuses on digital design and fabrication, with an emphasis on issues of modularity, sustainability, and tectonics. His work has been funded by the American Institute of Architects and the Austrian Government.

FRANCA TRUBIANO

Franca Trubiano is a registered architect, researcher and educator with research areas in construction technology, materials, tectonics, integrated design, architectural ecologies, and high performance buildings. She is editor and co-author of the recently published *Design and Construction of High Performance Homes: Building Envelopes, Renewable Energies and Integrated Practice* (Routledge

Press) and a Principal Investigator with the *Energy Efficient Buildings HUB*, a US Energy Innovation Hub. Her funded research is focused on Integrated Design Practices for Energy Efficient Building Retrofits and on Building Information Modeling (BIM).

DAVID YOCUM

David Yocum, AIA is Professor of Practice in Architecture at the Georgia Institute of Technology and a founding principal of the architectural practice BLDGS. He received a Master of Architecture from Harvard University, and a Bachelor of Arts from Dartmouth College with majors in History and Studio Art, where he was awarded a James B. Reynolds Fellowship. Recent work of the firm includes a conservative Jewish synagogue, private gallery and museum spaces for contemporary art, and university arts and classroom buildings.

Introduction

In the past two decades economic bubbles inflated and architectural spending around the globe reached a fever pitch. In both well-established centers of capital accumulation and far-flung locales heretofore seldom uttered in the same breath as the name of any Pritzker Prize winner, audacious building projects sprang up like mushrooms after a good rain. At the same time, the skyscraper, heretofore more commonly associated with the hurly-burly of American capitalism seemed only a few years ago as if it might pack up and relocate permanently from Chicago and New York and settle instead in Dubai and Shanghai.

Of course, much has changed since the arrival of the Great Recession. In formerly free-spending Dubai the tallest building in the world is now is named after the president of Abu Dhabi after he stepped in with last-minute debt financing. In cities across the United States housing prices nose-dived and freshly scraped lots sit ready for commercial redevelopment that likely has been setback by a decade. Similar stories are not hard to find around the world. The growth in Asia and the Middle East that many firms in the US and Europe counted on to pay the bills hasn't been sustained. Architecture firms that swelled in flush days are slimming down on crash diets and in the past five years jettisoned employees at a startling rate. The waves of capital and credit that once crashed on shores around the world are now at a low tide not seen in decades.

Although other professions have suffered painful reversals in the past half-decade, the decline in employment for architects was particularly precipitous. In the United States, for instance, the American Institute of Architects announced in late 2012 that firm revenues declined by 40 percent between 2008 and 2011. Employment at architecture firms took a similar hit, declining by nearly a third between 2007 and 2011. The decline in employment is especially stark, given in the boom years of 2003–2007 architecture firms grew, but only by 18 percent. The Great Recession wiped out the impact of the boom years and then some in the profession of architecture. The discipline of architecture (i.e. the academy) was somewhat insulated from the immediate impact of the contraction of the profession, as students already in school often stayed on to attend graduate school rather than enter an eviscerated job market with just an undergraduate degree. Many architects also saw the first years of the downturn as an opportunity to return to school. As the Great Recession dragged on, however, and the prospects of a career in architecture grew increasingly grim, the numbers of students choosing to study architecture inevitably declined.

Both the practice and the discipline of architecture face an economic, social, and political landscape vastly different from the one that greeted them in the early years of the 21st century. For the firms operating in the *starchitect* universe, life remains pretty good, and pretty profitable. A recession's paradoxical effect of increasing the purchasing power of the wealthy means that there are still plenty of projects out their building super-luxury high-rises. The World Cup and Olympics must go on, recession or not, and so new, ever grander stadia must be built, in locations as football mad as Rio and as callow as Qatar. The customary tendency of global elites to converge on a shared set of locations (NYC, London, Dubai) and styles (glossy, placeless, heavily fenestrated) means that often the same firms that design the high-rise design the stadium (see Herzog & de Meuron or Zaha Hadid). One can travel from the floor-to-floor windows of an apartment in Milan to much the same in New York. The match viewed from the luxury box of one stadium likewise matches the experience at the next one.

For the firms operating outside of the club-class level of practice, however, matters remain far less settled. Sustainability emerged in the last decade as a driver for schools, firms, and clients. Indeed, the number of LEED AP certified personnel on firm payrolls doubled between 2009 and 2011, suggesting that firms valued sustainability credentials even in the midst of shrinking markets. Advancements in computational design, BIM, and the increased integration of architecture, engineering, and construction practices to streamline design and construction also drove much of the dialog about the future of architecture during both the fat and lean years of the past decade. But none of these trends, encouraging as they might be in their own way, offers an answer to just how the discipline and practice of architecture will respond to shrinking numbers. Does the present moment call for a period of examination, reflection, and change? Or is it the case, as some argue, that once the economy rebounds (now, next year, five years from now) architecture's ship will be righted and service as usual will resume. Perhaps, but recessions rightly make people skeptical about the veracity of claims that rising tides lift all boats, and certainly the glacial pace of economic growth since 2008 doesn't inspire much confidence in the "wait just a bit longer" approach to finding solutions.

This edited volume brings together the work of scholars and architects proposing divergent and self-reflexive directions for the discipline and practice of architecture in light of the present moment of economic instability (and its attendant social and political consequences). Chapters address a range of questions: What sorts of programs (new or alterations of existing ones) might emerge from this moment? How might our thinking about materiality change when scarcity rather than opulence is the operative norm? What becomes of our tendency towards technological fetishism? Around which uses and causes will we develop new monumental or modest forms? How has the rhetoric of security (both economic and physical) come to define architecture in an age of recession and terrorism? How might eco-centric architecture develop in the midst of an economic contraction? Who will be the clients for any new modes of architectural thinking and practice?

These chapters thoughtfully critique the professional and disciplinary conditions nurtured in the past twenty years of illusory and lopsided prosperity and in turn suggest new and alternate directions for the academy and practice. If the crumbling of gilded ages in the past afforded architecture a moment of self-reflection, then this volume offers answers to the question of just what architecture could become in this present age of uncertainty.

1

Architecture in an Age of Uncertainty: Tales from the Recent Architectural Past

Benjamin Flowers

"There is a technical term economists like to use for behavior like this. Unbelievable chutzpah."[1]

How does one make sense of the last decade and a half of architectural production? It is an era characterized by a series of seemingly paradigm shifting experiences: 9/11, ecological disaster on a growing scale, the explosion of architectural projects in places expected and otherwise (Kazakhstan anyone?), and capped by a global recession second in severity only to the Great Depression (and one that 5+ years on shows no sign of abating, with Europe recently slipping back into economic contraction). Financial speculation, a seemingly endless deluge of petro dollars, and vast sums of capital from points East and West seeking physical expression spawned vast building projects and even entirely new urban landscapes around the world.

Perhaps one way to assess the state of architecture in our present condition of uncertainty is to look at some of the signal projects of this recent era. How did clients and architects respond to the grand challenges of the time? What overarching formal and expressive characteristics took hold? Which individuals, nation-states, and corporations were the wealthy patrons of this era and what building types did their patronage favor? In sum, how did the social practice of architecture respond to and likewise shape our understanding of an era of unprecedented change?

We could start with one project in order to begin answering the above questions: Herzog & de Meuron's 1111 Lincoln Road parking garage in Miami. Set aside the formal curiosity of the return of Brutalism after decades of public hostility to the style—the real point of interest here is the deployment of 10,000-watt star-architecture luminosity to sex-up a common urban eyesore. It seems at first like a kind of generosity—rather than gazing at the standard pre-cast slabs of concrete canted at angles to ease drainage and with head clearances of less than a foot you get a muscular structure—"all bodybuilder without the cloth" according to Jacques. It is such a spectacle that the upper floors are let for weddings and parties and on Saturday nights people with fabulous wheels vie to park their rides there.

But the heart of the matter is utterly non-architectural—the developer of 1111 Lincoln built the parking garage in order to secure approval for construction of the retail space at the ground level. Architecture here was a means, rather than an end. This is instructive for our survey.

A second informative project is the recently completed Moshe Safdie design for the Crystal Bridges Art Museum in Bentonville, AK. Cast as an act of public altruism ("free admission!") by its benefactor, Wal-Mart heiress Alice Walton (estimated net-worth: $23 billion), the structure (whose cost has not been made public) cloaks the origins of the capital that built it in a veneer of high art and cultural respectability. Poke a bit further, however, and the morphed simple geometries so characteristic of Safdie's design vocabulary lead to far less architecturally auspicious venues like the Tazreen Fashions and Rana Plaza factory buildings in the garment district of Dhaka, Bangladesh. Both produced cheap clothing for Wal-Mart (and other discount retailers). In 2012 a fire broke out at Tazreen Fashions, killing 112 workers. Many died behind metal gates at staircases the factory owners kept locked to control the movement of workers. A century earlier similar conditions in New York City were responsible for the death of 146 workers in a fire at the Triangle Shirtwaist Factory. In 2013 the Rana Plaza factory collapsed, killing 1,129 garment workers (mostly women) earning an average of $38 per month.[2] These (as-yet unprosecuted) acts of international corporate malfeasance are a reminder that Crystal Bridges, far from being an act of benevolence, is a poor effort at atonement from a family and corporation that were and remain global leaders in the development of business models (widely mimicked by other multi-national enterprises) dependent on extracting vast amounts of wealth from the surplus labor and lives of third world workers and mean wages of working-class consumers everywhere.

Of course guilt and insecurity have long been hallmark motivations for the construction of great architecture. Anxieties about class status and image motivated many of the clients who hired Richard Morris Hunt in the late 19th century. Hunt built his "cottages" (as he called them with classic understatement so popular among the outrageously wealthy) not to his client's specifications. Rather they came to him so that he might dress them in the language of distinction of which architects are understood to be arbiters. As Robert Hughes aptly noted: "At all times in America, the poor have wondered how the rich live. But more to the point, the rich have always wondered too." It was R.M. Hunt who showed the rich how they should live. In the twentieth century, the architect continued to provide clients with desired social status. The clients for the Empire State Building and the Seagram Building—two of the great NYC skyscrapers—were motivated by their exclusion from the upper echelons of society (due to religious and class status) to build tall. It is also the case that on occasion, building for clients of colossal wealth and power is part of an architect's quest for status (and redemption). Such was the case with Minoru Yamaskaki and his work on the World Trade Center (coming as it did while his early work at Pruitt-Igoe was being dynamited).

An exchange of crass cash for laundered elite cultural capital also animated building across the globe over the past 15 years. Tyrants, economic, political, and otherwise, turned to designers to provide them buildings that spoke of legitimacy, modernity, and all the values associated with an appreciation for fine architecture. If one were to map the congruence of human rights with ambitious architecture, a curiously inverse relationship emerged in this period. A City-state imported indentured servants paid $4 per day to construct the tallest building in the world. Nations that criminalize the full expression of a person's humanity (gender rights, the freedom to declare love to whom one chooses, the freedom to avoid genital mutilation) brought the necessary amount of capital to the table and were rewarded with the architecture for which they paid. An equally curious metric could be drawn on the relationship of corruption to grand building. The corruption index and the building index walked hand-in-hand. We witnessed an era of design increasingly yoked to the absence of freedom and democracy rather than its presence. At one time modern architects imagined their task as implicitly political. Describing the ambitions of his generation the architect, writer, and design journal editor Peter Blake wrote: "we believed that we could slay the automobile, defeat fascism, and abolish disease."[3] What would he make of 1111 Lincoln and its submission to the tyranny of the internal combustion engine? By the start of the 21st century, however, it was clear that similar ambitions rarely held for many Pritzker Prize winners. As one architect on audacious projects for wealthy but suspect clients told me, "it is just a business."

The architecture of the past 15 years illustrates an important but often unstated reality: it is never a bad time to be rich. The bubble economies of the late 1990s and early 2000s produced a range of projects satisfying the yearnings of the newly rich and the long-since rich. When the recession came about, the fictive-rich found themselves in trouble, and projects tied to them did indeed wither on the vine. But for the wealthy, the recession only added to their buying power, as global demand for building materials and labor declined, and declining real estate values overall made investment in that sector increasingly desirable. Thus, even during a period of economic decline and contraction –not to mention vicious cuts in public spending—record-setting real estate deals closed in cosmopolitan metropolises around the world. In one city, New York, the real estate market is now fundamentally deformed by the declining fortunes of the strangled middle and working classes (who can afford very little) and the vastly rising wealth of the super-rich (who can afford anything). There are at present 49 buildings in New York City with apartments selling for $15 million or more. That is a nearly 50 percent increase since 2009. Plans for some 20 additional projects are in the planning stages or under construction. One building, 432 Park Avenue, has a penthouse that sold for $95 million, as well as 4 and 3 bedroom units selling for $45 million and $32 million respectively. The architect of that project, Rafael Viñoly, notes that there are now only two markets in New York City: "ultraluxury and subsidized housing."[4] Total sales at 432 Park are expected to surpass $3 billion.

At 15 Central Park West, designed by Robert A.M. Stern, total sales exceed $2 billion. Aside from providing an index of the canyon-like gap in wealth in NYC (where median family income has declined by 8 percent since 2008), these buildings operate as commentaries on the mindset of economic oligarchs. Much has been written about the role of the banking and financial services sector in eviscerating the global economy. An examination of sales records of 15 CPW in 2005–2006 offered a warning flare (little heeded at the time) that bankers and financiers lacked the basic competence to be trusted as stewards of our collective economic well-being. It was that year Sandy Weill, CEO of Citigroup, paid a then-record price for his unit in the building: $XX million. At $XXX per square foot, this was a price utterly unmoored from any reasonable economic or geographical reality. The real estate pages heralded the sale, as expected from the booster arm of the fourth estate. But the business pages should have been able to connect the dots and realize that anyone willing to over-pay so egregiously for an apartment could not possibly discern between a clever credit default swap and a toxic asset in disguise. As one broker whose record of trading was profitable before it was revealed to be fraudulent put it about his bank's executives: "they never dared ask me any basic questions, since they were afraid of looking stupid about not understanding futures and options."[5] Between 2005–2006 the hedge fund executive Daniel Loeb bought an 8-bedroom unit for $5 million (or $4200/sq. foot) and Lloyd Blankenfein, CEO of Goldman Sachs, bought an apartment for $26 million, suggesting that Weill was not an outlier among financiers in his irrational exuberance.

It is more than the financial absurdity of these real estate prices that offer a glimpse into the mindset of the titans of Wall Street and global capitalism. There are the aesthetics of these projects to consider—these are after all buildings intended to help the rich understand themselves, Richard Morris Hunt-style. At 15 CPW, the great and powerful are sheltered in a building whose massing recalls the setback buildings of Candela. The structure is divided into two parts, a lower, symmetrical 19-story portion referred to in the casual understatement of the truly wealthy as "the house." Rising behind the house is a tower of 43 stories whose final 10 or so stories are disposed asymmetrically. A formal court that runs the length of the site divides the two units. Amenities include a private driveway to elude photographers. Formally conservative, and with an air intended to stroke the egos of residents and impugn the standing of passers-by, the whole structure is clad in 85,000 pieces of limestone. It is a building whose ambitions are almost entirely financial, with little about its formal characteristics to describe as novel, innovative, generous (in the public sense), or noteworthy (perhaps beyond the scale of the entry lobby). It is a building that embraces monumentality for its own sake unambiguously and without hesitation. This is a building for people who wish to be reassured that no matter what changes in this world, they will still occupy its uppermost echelons. It feels a bit like one of the Seven Sisters has been repurposed for a new generation of tyrants, and so perhaps it is not so surprising that when Sandy Weill decided to sell, he did so to a Russian billionaire, Dmitry Rybolovlev. Rybolovlev, who made his fortune in the fertilizer industry after the break-up of the Soviet Union, paid $88 million for the apartment. A titan of finance sells to a titan of fertilizer—it is hard to know which business is more full of it.

Where 15 CPW strives for a nostalgic echo the work of Candela and the era of the setback tall building (and attendant Gatsby-like era of excess), 423 Park employs a divergent formal strategy, opting for an organizing system of rectangles. The building has a square 93 by 93 foot base, with each façade of the 84-story tower divided by six 10 by 10 foot windows. It is a pattern language of wealth and authority somewhat less accessible to the general public than Stern's limestone bacchanal, and therefore requires contextualization in order to be marketed to the target audience of nomadic globetrotting billionaires. To that end, Harry B. Macklowe, the developer, spent over $1 million on a four-minute promotional film for 432 Park Avenue. As reported in the *New York Times*, "the movie consists of a series of dreamlike sequences, rife with images of wealth and privilege, and loosely plotted around a stunning brunette as she travels from her home at 432 Park Avenue from her country Estate in England. She is shown leaving the manor in the backseat of a 1957 Rolls Royce and then flying across the Atlantic in her Learjet." Later in the film, Philippe Petit, who famously tightrope walked between the World Trade Center towers in 1974, is shown walking a tightrope from the Empire State Building to the tip of 432 Park Avenue. Canonical architecture is yoked to the task of promotion: the film compares the building's lap pool to Mies' Barcelona Pavilion and its gridded façade to the ceiling of the Pantheon.[6] The film includes a party at which the mysteriously resurrected specters of Le Corbusier and Al Capone, of course, dance with Harry Macklowe wearing a King Kong costume.

It feels all like someone is trying just a bit too hard to make out hyper-expensive speculative real estate ventures (both their creation and their consumption) as something fun. The self-doubt underlying the need to compare a lap pool to the Barcelona Pavilion is particularly revealing. Mies was no stranger to luxury, or vast architectural expenditures, but anyone who has spent some time in Houses Farnsworth and Tugendhat or the Seagram Building knows that this whatever else Miesian luxury is about it's not *fun*. Vinoly's building, like Stern's, is invested in references to work from the 1920s, only Vinoly is looking to Italy rather than New York. The insistence on a single organizing geometry—the square—disposed in white recalls (albeit at a smaller scale) Terragni and his civic buildings for the Fascist government. For much of the post-WWII era architects struggled with monumentality, as in the wake of Fascism in Germany and Italy, and pushed by the heightened political tensions of the Cold War, monumental architecture necessitated a robust socio-political framing. Otherwise, one risked having a monumental structure's sensitivity *mis-read*. Two decades after the fall of the Berlin Wall, architects and clients no longer seem to require that their monumental buildings speak in with such precision. The result is a moment in which the super-rich purchase both nostalgia and discarded ideologies in the same time and same city. The inherent anti-individual and seemingly inhumane dimensions (10×10) of the window aperture at 432 Park Avenue might have been cause for pause in an earlier era. Those dimensions might fit in a department store façade at street level or perhaps an office tower, but would hardly be thought scaled to the demands of domesticity much less the dimensions of an individual. After all, as we were in the west told time and time again, it was capitalist democracy that favored the heroic individual and communism that sought to devalue the same. Today, with tall buildings scrubbed of their need to situate themselves within any framework other than that of consumption, no one seems to much care that the tallest residential building in the western hemisphere looks like a monument to faceless, nameless, unknowable, and endlessly repeated authority.

The impression 432 Park gives of authority unmoored from the regulatory regimes that guide and confine the aspirations of the mass of men and women who build, maintain, serve, and clean the city around it is no accident. While working-class and middle-class residents of New York and other major cities in the United States struggle with an unprecedented wave of foreclosures and declining real estate values, the ultra-luxury market is the recipient of government largesse in the form of tax subsidies. One57—another tower residential project, this one designed by Pritzker Prize winning architect Christian dePortzamparc—has apartments under contract for $90 million. But it also received at least $50 million in tax credits under a program designed to encourage—yes—affordable housing.[7] The impact of the purchasing power of the ultra-rich extends beyond the ultra-luxury market. Real estate is the largest asset class in the world, and this means a variety of institutional actors (real estate investment trusts, hedge funds, and other arms of investment equity) are buying housing across the nation, often in markets hit the hardest by the financial crisis of 2008. The hedge fund managers living in $50 million dollar apartments in NYC are also buying tens of thousands of houses at a time in Florida and California. Just two examples are Blackstone Capital, who recently bought 26,000 homes and Colony Capital, which owns 10,000. In Florida and California in 2012 investors, many of them large firms and many paying in cash, made 35 percent and 25 percent respectively of all home purchases. The paradox is that these investors are pricing out private actors—individuals looking to buy housing primarily as residences rather than investments.[8] Thus, although it might be tempting to dismiss the high-end apartment bubble in NYC as an isolated condition, the capital supporting those sales prices is directly linked to the financialization of housing markets across the nation, affecting cities and states far removed from New York City and unlikely to ever have a Vinoly or de Portzamparc project of their own.

The scope and breadth of the financial and regulatory manipulation at the core of these NYC projects is replicated at other scales and other locales. Rather than isolated examples or outliers, the cozy relationship of architecture to fundamentally inequitable economic actors is the norm for the past decade and a half. We've already touched on Crystal Bridges, an edifice completed less than 18 months before the collapse of a factory in Bangladesh producing clothing for Wal-Mart. Two corporate headquarters projects underway in California are worth noting: Sir Norman Foster's design for Apple and Frank Gehry's for Facebook. Although Apple has been around longer than Facebook, the two companies experienced explosive growth over the past 15 years as product design and social media established themselves as lynchpins of some new economy. Both companies are at present astonishingly ubiquitous. The decades after the end of World War II were boom-year eras in the construction of corporate headquarters for companies involved in the expanding consumer and service economies (Seagram, Pepsi-Cola, John Deere, Pan Am, etc.). The prevailing wisdom was that these new buildings would help define a modern, positive, and stable corporate identity for these increasingly large, multi-national corporations, one that consumers would then attach to the goods and services these companies sold. This made sense in an era when consumers were faced with increasingly large and anonymous corporate identities that deviated significantly from their pre-war, Depression-era antecedents. But neither Apple nor Facebook fit that client model. Both are companies with robust public identities, indeed ones closely tied to the charismatic traits of individual executives: Steve Jobs and Mark Zuckerberg. Regardless of whether you like either one of them, it is certainly the case that both are far better known than the executives in charge of any of the postwar corporations mentioned earlier. Thus, it would seem that neither company requires the luster of capital-A architecture to enhance their already well-established reputation. After all, can a single building move the needle of public perception for either of these firms? Not likely. Why then build, particularly at the price point these two corporations are? Apple's HQ is expected to cost over $1 billion, a hefty price tag even for a company currently sitting on nearly $1.5 billion in cash reserves.

The answer is that as the wealthy and powerful grow more so, so too is there an increase in the perceived power of architecture to cloak that wealth and power. But this power accrues to an ever-smaller coterie of architects and design firms. In essence these architects serve as the design equivalent in "in-house" counsel—they jettison their own ethical scruples, political principles, or social values, and instead agree to serve the interests of the client above all else. As architecture is fundamentally a social practice, this represents a radical narrowing of responsibility on the part of the architect. Paradoxically, it speaks to the esteem these wealthy and powerful clients attach to architecture (perhaps odd, given that many of these clients attach seemingly so little esteem to other customs and laws).

Indeed, the degree of disdain for the custom of good corporate citizenship is especially telling. Both Facebook and Apple have well-established (if only recently made clear to the public) practices of engaging in spectacularly creative tax liability accounting. Between 2009 and 2011, "a web of subsidiaries so complex it spanned continents and went beyond anything most experts had ever seen" helped the company avoid—with the aid of schemes known as the double Irish with a Dutch sandwich—paying over $8 billion in taxes in the United States, making the it, in the words of one US Senator, "among America's largest tax avoiders." Although Apple is happy to label all of its products and advertising with the phrase "Designed by Apple in California," for tax purposes it claims to be "a resident of nowhere."[9] Facebook likewise employs a variety of complex strategies (including the double Irish) to make the argument that although their operations are based in the physical world its profits (shuttled to Ireland and then the Cayman Islands) are stateless nomads.[10]

Elsewhere around the world matters of architecture were likewise strange. Zaha Hadid designed grand and cosmopolitan projects seemingly everywhere: the Rosenthal Center for Contemporary Art (Cincinnati, 2003), Phaeno Science Center (Wolfsburg, 2005), MAXXI (Rome, 2010), Opera House (Guangzhou, 2010), Riverside Museum (Glasgow, 2011), and the Broad Art Museum (East Lansing, MI, 2012). And that is just a partial list. One could imagine that an architect of her stature would be increasingly particular about selecting clients in this moment of grand ascent in her career. And yet, situated among Soviet era apartment blocks in Baku, Azerbaijan is

Hadid's Heydar Aliyev Cultural Centre (2013). Named after the ex-KGB general and autocrat who ruled the nation for nearly 40 years before handing over power to his son and then dying of stomach cancer, the characteristically Hadid structure (lots of gloss, a lot less context) is an outlier for Baku, but not for starchitects.

Indeed the recent past was witness to seismic shifts in wealth production and its attendant architectural dress, and many big-name western firms seemed unperturbed to operate on behalf of leaders and nation-states with no pretense of democratic aspirations. Bjarke Ingles designed the national library in Astana, Kazhkstan for strongman president Nursultan Nazerbayev, even cleverly Photoshopping the long-time ruler into a rendering of the interior hall in the proposal renderings. Nazerbayev also employed Sir Norman Foster to dot Astana (the built-from-scratch capital city erected via decree) with a pyramidal Palace of Peace and Reconciliation and a neon-lit Entertainment Center that also is the world's largest tensile structure. Many in the glow of knighthood might have resisted the charms of Astana, but not Foster.

As grand as Nazerbayev's ambitions for building are, they are dwarfed by his neighbors to the south. For half a century the tallest buildings in the world were found in New York City, by 2010 it was Dubai that was home to the tallest—the Burj Khalifa. A curious inversion accompanied this geographic shift. The skyscraper emerged from the confluence of new building technologies (steel framing and the elevator for instance) and rising demand for a finite resource (real estate). Defined by Cass Gilbert succinctly as "a machine to make the land pay," the tall building is generally understood as a symbol of wealth and economic boom times. Of course, the tallest building in the world for nearly 40 years, the Empire State Building, opened in the midst of the Great Depression and didn't turn a profit for nearly two decades. The Burj Khalifa, a $1.5 billion project, likewise is a model of economic non-sense. Originally called the Burj Dubai, the building was renamed after capital from a neighboring Abu Dhabi ruler saved the day when the building boom collapsed in Dubai. Over 160 stories tall, it holds offices, apartments, an Armani-branded hotel, along with several sky lobbies and restaurants. The irrational exuberance of Dubai's building spree is particularly noteworthy as the labor force that built it and maintains it is almost exclusively foreign and paid wages that would not have been out of place in the Great Depression. Skilled carpenters working on the Burj Khalifa were paid less than five pounds a day; less skilled labor made less than three pounds a day.

The boom of the 2000s in the UAE generated more than multi-use supertall buildings—it also brought outposts of well-established western cultural institutions. The Louvre and the Guggenheim both announced plans to open branches (designed by Jean Nouvel and Frank Gehry respectively) on Saadiyat Island in Abu Dhabi. These will sit nicely next to projects by Zaha Hadid (Performing Arts Center) and Sir Norman Foster (Zayed National Museum) in a cultural district built from the sand up. The aura of fashionable cosmopolitan glamour these museums are intended to project on a global stage is difficult to reconcile with news reports documenting the strict limits on personal freedom and safety imposed by the legal system in much of the UAE.

Perhaps the capstone moment of the this-is-not-what-it-appears-to-be era took place in China, which too went through a massive building boom. In Shanghai and Beijing and other cities across China architects spent the last decade and a half building the new, urbanized Chinese state its leaders envisioned. Capital-A architects were brought in and audacious buildings rose in great numbers. Among the most famous of these was the Office for Metropolitan Architecture's design for the headquarters of CCTV (China's state broadcaster) in Beijing. Radically cantilevered and clad in a high-tech façade, the headquarters were affectionately nicknamed "the pants" by locals. (A building designed by RMJM in Suzhou was also called "the pants," but in this case as a criticism. Apparently not all pants are equal.) In 2009, as the effects of the global financial markets collapse were being felt, CCTV employees celebrating the Lunar New Year set off illegal fireworks that ignited a fire at a neighboring hotel (also designed by OMA), destroying it. Although the event was widely captured in cell phone images, CCTV news failed to cover the inferno raging next door. It was left instead to Chinese micro-bloggers to spread news of the event. One who did posted a image of the hotel in flames next to CCTV with the caption: "I am not scared, the fire has not spread to the big underpants yet!" An overheated real estate boom literally burned out in spectacular fashion. What lessons we will draw from these calamitous years and the subsequent great recession remains to be seen.

NOTES

1   Edward Kleinbard, quoted in the *New York Times* (26 May 2013), D2.

2   "Justice Still Elusive in Factory Disasters in Bangladesh," *New York Times* (29 June 2013).

3   Peter Blake, *No Place Like Utopia: Modern Architecture and the Company We Kept* (New York: Knopf, 1993), 9.

4   "Sky High and Going Up Fast: Luxury Towers Take New York," *New York Times* (19 May 2013), A1, A16.

5   Floyd Norris, "Masked by Gibberish, Risks Run Amok," *New York Times* (21 March 2013).

6   "'Dream a Little Dream,' at $250,000 a Minute," *New York Times* (23 June 2013), BU 11

7   "Luxe Builders Chase Dreams of Tax Breaks," *New York Times* (25 June 2013), A18.

8   "Behind the Rise in House Prices, Wall Street Buyers," *New York Times* (4 June 2013), A1-A3.

9   "Apple's Web of Tax Shelters Saved It Billions, Panel Finds," *New York Times* (20 May 2013). See also "Apple's Move Keeps Profits Out of Reach of Taxes," *New York Times* (2 May 2013), "The Corrosive Effect of Apple's Tax Avoidance," *New York Times* (23 May 2013), "The Trouble With Taxing Corporations," *New York Times* (28 May 2013).

10   "Facebook Paid £2.9m Tax on £840m in Profits Made Outside the US, Figures Show," *The Guardian* (23 December 2012).

Architecture and the Vicissitudes of Capitalism

William Mangold

1

In an age of uncertainty we ask, "What is important?" The spaces and forms that make up our built environment provide us not only with the basic requirement of shelter but also with an endless array of experiences. As we consider what is meaningful in our lives, certainly our experience of architecture—our city streets, the places we call home, our edifices of culture—ranks high on the list of things we value. However, as with most things, our relationship to architecture is complicated, especially in our milieu of capitalist production and consumption. This essay will explore some of those complications, as interpreted through a Marxist lens, and argue that capitalism has exploited architecture, leaving us with only a shell of what could be a rich and fulfilling experience of the built environment. Within this unhappy picture emerge a few bright spots and possible directions through which architecture could be redeemed.

2

To begin, there are at least three ways of thinking about what architecture is. The first understands architecture as the buildings that make up our environment. There are arguments about what can be classified as "architecture"—which structures are Architecture versus which are mere buildings—but this definition fundamentally describes an identifiable built *product*. A second definition, probably more accurate, recognizes architecture as a *process*. This takes into account the work of architects to produce drawings, which are made into buildings by the construction industry. In this case, "architecture" is the work done by architects in designing and overseeing construction. Again, there may be argument about when that process begins and ends, but it can be distinguished from the buildings that may result from the process. A third way of thinking about architecture would be in terms of *production*. This definition could include a number of activities—such as education, publishing, and exhibiting—that accompany the making of buildings and are often carried out by people that are not architects. It also could include a number of products—books, drawings, renderings, models, websites—that are not buildings. These different ways of understanding architecture begin to hint at how the role of architects and the things that are produced may be open to exploitation.

Turning briefly now to Marx, we will consider his descriptions of exploitation before examining how it can be understood in connection with architecture. There are two forms of exploitation that I will look at here. The first is the exploitation of labor and the second is the exploitation of value. Regarding the exploitation of labor, in his *1844 Manuscripts*, Marx writes:

> *With the increasing value of the world of things, proceeds in direct proportion the devaluation of the world of men. Labour produces not only commodities; it produces itself and the worker as a commodity—and does so in the proportion in which it produces commodities generally. This fact expresses merely that the object which labour produces—labour's product—confronts it as something alien, as a power independent of the producer. The product of labour is labour which has been congealed in an object, which has become material: it is the objectification of labour. Labour's realization is its objectification. In the conditions dealt with by political economy this realization of labour appears as loss of reality for the workers; objectification as loss of the object and object-bondage; appropriation as estrangement, as alienation.*[1]

This passage outlines what Marx goes on to describe as the exploitation or "alienation" of the worker through the transformation of his labor power into an object. The worker is removed or "estranged" from what was most his—his own productive capacity. His lifeblood has gone into an object that is no longer his.[2] Marx is explicit and detailed in his discussion of how people with the capital necessary to employ labor are able to extract value through the production process. By turning labor power into sellable commodities—and paying less for labor than the value of the commodities produced—capitalists are able generate a profit. In this form of exploitation, we give up our labor power to produce a commodity that we do not own, but which sits before us and we are compelled to obtain.

According to Marx, exploitation also takes another form in which truth is divorced from reality. This comes in the divergence of exchange-value from use-value. Marx introduces these terms in the first section of *Capital* in his discussion of commodities. Use-value indicates the objective amount of labor-power put into an item, whereas exchange-value is a subjective amount established through social interaction. The divergence of these two forms of value has a couple of consequences. The first is that commodities may be exchanged at a different rate from what their use-value would indicate—which also serves to obscure the value of labor. A second consequence, which greatly concerned Marx, was that exchange-value, in the form of commodities, would entice production of useful articles solely for the purpose of exchange. "This division of a product into a useful thing and a value becomes practically important, only when exchange has acquired such an extension that useful articles are produced for the purpose of being exchanged, and their character as values has therefore to be taken into account, beforehand, during production."[3] This has further consequences: first that production is modified to result in greater exchange value, and second that it requires labor to both satisfy a social want (i.e. have use-value) and be mutually exchangeable (i.e. have exchange-value). Marx also remarks that these two facets of value have the consequence of making value a "social hieroglyphic" that becomes impossible to decipher. It compels us to ask why are things being produced—because they are useful, or because they can be sold?

3

With value impossible to decipher and labor-power estranged from the worker, the door is opened for exploitation. Capitalism is the chief vehicle of exploitation today. While I will not go further into the mechanisms of capitalism in this chapter, suffice to say that it employs the two strategies outlined above (exploitation of labor and exploitation of value) to accumulate capital in the form of money. Although Marx does not specifically address architecture, it is now possible to consider a number of interconnected ways in which architecture is exploited.

COMMODIFICATION

As Marx suggests, when something becomes a commodity it is on the path to exploitation. This is certainly true for architecture in a variety of ways. In its built form, architecture is commodified when it is bought and sold, or discussed in square feet and number of bathrooms. This is also true at the scale of materials. When a tree is cut down and sold as lumber or clay is made into bricks, the earth is exploited and turned into commodities. As commodities, materials and buildings are exchangeable and become principally thought of in terms of quantified exchange-value rather than for the quality of their use-value. Real estate developers rarely deliberate on the feelings or experiences someone may have, instead they focus on maximizing the value of every inch. In the same way, the process of architecture can be commodified as the services of architects and designers are measured in hours and productivity. Designers become labor-commodities in the building process as their services are measured for the exchange-value they contribute to a project. This is no truer than in the employment of "starchitects" on contemporary projects to exploit the name brand of certain designers to increase the exchange-value of buildings they work on. One example is the faux-classical building by Robert Stern at 15 Central Park West in Manhattan, which at approximately $2 billion in sales ranks as the highest-priced new apartment building in the history of New York.[4] While location and distinction are major factors, his name also adds cachet. These instances that characterize the commodification of architecture lay the groundwork for other means of exploitation.

GENTRIFICATION AND URBAN "DEVELOPMENT"

When buildings are exchanged as commodities, investors and developers play the market in search of profit. This leads to a cycle of real estate investment and dis-investment. Neil Smith identifies this as the pattern of gentrification.[5] Smith describes gentrification as a cycle that begins with periods of dis-investment during which buildings and neighborhoods are intentionally neglected by capital in order to drive down their value. Real estate prices fall, conditions further decline, until these areas can be re-conquered by pioneering artists and designers looking for cheap space. For property owners and developers, the interest by the design vanguard signals an opportunity to re-invest, improving the quality of the neighborhood and opening it to more mainstream residents. This process, Smith explains, is underpinned by the practices of financial institutions, as well as the policies and operations of city governments.[6] Capitalism drives the process of urban "development" to manipulate land values in order to extract a profit from real estate investment. In this pattern of gentrification, artists and designers are exploited and buildings and neighborhoods are held hostage to the profit motive of capitalism.

ABSORPTION OF SURPLUS CAPITAL

Another pattern that emerges when buildings are treated as commodities—perhaps even more sordid and pervasive—has been identified by David Harvey. "Spatial fix" is the term he has used to describe how capitalism uses urban development as a locus for surplus capital, in order to avoid crises of overaccumulation. Since Haussmann's activities to transform Paris, buildings, real estate, and infrastructure have been increasingly used to absorb surplus capital.[7] While at first glance this might appear beneficial to architecture—ready capital allows for heightened architectural development—this activity becomes unsustainable as capital seeks its profit. What seemed to be good for architecture turns out to be a thin mask for capitalist exploitation. By absorbing the surplus capital, buildings and infrastructure provided a safe reservoir to offset the faster-paced cycles of commodity production, but as profit is eventually extracted, capital leaves behind cheap, shoddy buildings and sucker-homeowners holding the bag. Of course this is further beneficial to capitalism through the cycle of dis-investment and gentrification. Real estate deteriorates and the next wave of creative destruction begins.

SOCIAL REPRODUCTION AND QUALITIES OF HOME

We all get used to a low-quality built environment and accept it as the norm. Our acceptance of commodified buildings, cheap construction, and the exploitation of design occurs through a process known as social reproduction. Social reproduction is a complex and dynamic process, but there are a few examples of how norms are established and reinforced that are worth discussing in regards to architecture. One of the ways in which social patterns are established is through the production of desire.[8] In the realm of architecture and design, desire is produced through ubiquitous media such as home remodeling television shows and images circulated in print. What may begin as desire is reinforced by the limited options people are given when it comes to the built environment. People are induced to consume building products, but their choices are severely limited by standardization and mass production, which again is driven by capitalist profit seeking. Everyone from manufacturers to retailers to construction contractors stand to benefit from offering fewer options and charging a premium for customization. Although architecture has the potential to be uniquely adapted to the needs and conditions of its inhabitants, the demand for profit often forces consumers into a generic box.

These examples describe the productive and consumptive aspects that shape social reproduction, but there are other processes that reinforce social norms in deeper and subtler ways. Witold Rybczynski discusses how we have slowly come to our contemporary notion of "home" in which everyone is expected to live within a private, individualized sphere.[9] The expansive possibilities of architecture are constrained by the assumption that every family home must have its kitchen, dining room, master bedroom and bathroom suite, and two-car garage. Even the most progressive design is reduced by the social expectations about what our spaces should offer.[10] This has resulted in stunted advancement of sustainable design options and continued limitations on the environmental choices of people that have disabilities.[11] Perhaps the most insidious issue is our continued construction of suburban McMansions. This type of residence persists as the emblem of middle class life in America as a consequence of spatial privatization and capitalist alienation. We continue to love and build these houses because our socially reproduced ideal remains the myth of individual liberty, in the face capitalist domination.

ARCHITECTURAL EDUCATION

On the surface, issues of social reproduction are what education seeks to mitigate—more educated designers (and clients) should be able to make better design choices. However, this problem of social reproduction is often exacerbated by education in architecture. Some schools of architecture intentionally reproduce the status quo and make no effort to challenge social convention or the forces of capitalism. They embrace the role designers play in the production process and the highest goal these programs strive for is gaining employment for their graduates. Other schools are unintentional reproducers—teaching a cannon of design that reinforces the norms without considering the consequences. However, most schools imagine themselves as progressive or even rebellious. Nevertheless, unable to modify social expectations or situate themselves to push back against capitalist hegemony, they fall back into an easy discourse about styles and formal aesthetics. These schools, while challenging the appearance of architecture, remain confined within the bounds of what is socially acceptable and expected. Unfortunately this fallback position of architectural education has a double-edged consequence that opens the door to further exploitation of architecture.

FORMALISM

When designers focus their energy on making eye-catching forms and debating how things look, they become marginalized in the production process. For one reason, formal arguments about style and aesthetics tend be seen as less important within the larger context of economics and production. These "aesthetic" debates can be written off as frivolous and secondary to practical concerns. At the same time, because design does not actually challenge or modify the practical concerns of how people inhabit places, it can be viewed as superfluous. The emphasis on aesthetic arguments and the failure to rethink the ways in which people live are indicative of a profession that has lost its sense of purpose. At best, design is merely employed to produce desire, but more often it becomes seen as irrelevant and a waste of money. Unfortunately this has further consequences that intensify exploitation in architecture. When design is perceived as unnecessary, this creates an atmosphere in which architects must constantly justify their services...and reduce their fees. This induces competition among architects and magnifies the degree to which they can be exploited.

SPECTACLE, "INNOVATION", AND "VALUE ENGINEERING"

Competition between designers further emphasizes style over what little substance might be possible. Architects race to produce images—"money shots"—that will grab attention, buildings are branded, and apartments are staged for sale. The visual is valued above the tactile—the spectacle over experience.[12] Emphasis is placed on "innovation" and novelty, which quickens the pace of production and consumption. No time is allowed for research or to develop projects thoroughly. Instead the process is streamlined, buildings are standardized, and perhaps in the most sinister twist of all, projects are "value engineered."[13] This term, perfectly descriptive of the process capitalists use to extract the most profit from their projects, confronts architects at every step and aptly summarizes the ways in which architecture is exploited.

4

I've painted a pretty wretched picture, but is it really all bad? Perhaps not. It is possible to tease out of this rough description of how architecture is exploited a few ways in which architecture may benefit from its engagement with capitalism. In periods of investment, architecture and the role of architects expands rapidly.

During these times, increased production may put greater demands on architects, but it has also meant more opportunities to build and a greater diversity of buildings constructed. Likewise, slow periods of building have often been attributed to the strongest growth in academic and conceptual development in architecture. This view holds that interaction with capital—during boom and bust—is good for architecture as it progresses as a discipline and profession.[14]

Another point of view suggests that the portrayal of architecture in the media indicates that there is a growing appreciation for design. As people are more exposed and become more aware of design, they are more likely to understand and desire to modify their environment. This would re-value design, making it worth the expense, and move it from the margins closer to the center—thereby reversing the pattern of social reproduction and competition described above. In this case, designers become instrumental and architecture plays a role in changing social norms.

A third way of thinking about the significance and sublime beauty of architecture and production is suggested by Walter Benjamin through his notion of phantasmagoria. In one of the most striking passages in his essay, "Paris, Capital of the Nineteenth Century," Benjamin introduces the concept to describe the experience of the Arcades in Paris—a fantastic architectural space intended as a marketplace for commodities.[15] Benjamin is keenly aware of the contradictions inherent in this vivid experience, but unlike some critics who dismiss it outright, Benjamin is drawn to explore this fascinating and dynamic realm of architecture and commodity. As critics like Marshall Berman point out, there is room for a similar approach today.[16] While it is possible to indicate the ways in which architecture is open to exploitation by capital, it is also possible to experience the heady and often remarkable constructions made possible through capital.

5

I would like to make a few more points before drawing to a conclusion. First, it should be clearly noted that architecture is not necessarily the innocent victim in the processes of exploitation I've described above. Architecture needs capital to be built and is often willing to do business with the capitalist devil in order to be realized. Architecture not only deals with capital out of necessity, but often architecture benefits—at least in the short term—from this relationship. As pointed out above, architecture expands and grows with capital, and in some cases architects themselves stand to profit through investment or real estate development. Another example of architecture's complicity with capital is presented by Anthony Ward, who argues that architecture frequently provides a screen for the operations of capitalism.[17] In his examples, he shows how discourses centered on form, function, or linguistics don't allow for discussion of the needs and people that architecture should serve. He posits that this deception—architecture focused away from people—creates a mask for capital to continue extracting profit, rather than provide for the inhabitants of these projects through a participatory process.

Designers must ask themselves, "Am I motivated by profit and cultural status, or am I committed to engaging and serving people who can truly benefit from a better built environment?" Too often the idealism of youth is transformed into the calculated operations of the profession. Professionalism should ensure a high set of standards for environmental quality and ethical practice, but more frequently it serves as an exclusionary mechanism that channels projects towards the dominant firms. Often, these firms have gained their positions of dominance through their willingness to service the needs of capitalism. If we are willing to ask questions about how designers can challenge the demands and conventions of a market-based economy, then it is worth discussing what can be done about our situation.

Perhaps the most immediate solution is for the profession to pursue a more local and participatory approach that better integrates the voices of users and communities. Designers can initiate projects to address issues they see in front of them,[18] and work with their clients to reduce focus on bottom line profitability. As for broader changes, some critics contend that architecture, as a profession, should be socialized—much like medicine in some parts of the world. The services of architects could be made publicly available and subsidized by the government. To some degree this is the recent situation in the Netherlands and was once a possibility in the U.S. in the 1930s. This arrangement would allow designers to address the needs of a far greater and more diverse population and could relieve architecture of the pressure of commodification.

Another possibility lies in Benjamin's exhortation from the "Author as Producer." Roughly paraphrased, Benjamin argues: "The more completely the architect can orient his work toward mediating activity to adapt the apparatus of production to the purposes of the proletarian revolution, the more correct the political tendency of his work will be, and necessarily also the higher its technical quality."[19] This proposition suggests that architects continue to practice, but at the same time begin to resist or subvert the demands of capital, and actively work to dismantle or adapt the system of production.[20] In so doing, Benjamin imagines that the architect could continue to produce a dynamic and phantasmagoric environment while shifting their effort toward the aims of revolutionary practices.

A final possibility for disengaging architecture from capitalism is through the intervention of mediating institutions. If organizations such as schools, museums, and not-for-profit design resources were able to buffer architects and the process of making buildings from the demands of capitalism, it could create a territory in which designers could develop projects and offer services that they would otherwise be unable to do. As suggested above, schools and other institutions are positioned to challenge conventions that are socially reproduced. Education, especially at the college level, is an opportunity to test alternatives and engage communities within the context of a stable and supportive system that is not expected to produce a profit. Projects by the Rural Studio at Auburn University or the Detroit Collaborative Design Center at the University of Detroit-Mercy come to mind as successful examples of how schools can provide a way for students to think differently about the possibilities of design.[21] Other organizations have also had success as platforms for designers to operate without the normal constraints of capitalist production. Van Alen Institute has historically held design competitions intended to generate ideas, stimulate conversation, and propel the work of imaginative young designers.[22] Design Corps and Architecture for Humanity[23] were founded in response to specific social and humanitarian needs, and have expanded to provide opportunities for designers to engage communities and develop architectural responses to crises all over the world. While none of these represents a silver bullet solution to the monstrous challenge of capitalism, they do suggest ways of re-working our current social and spatial conditions.

6

The relation between architecture and capitalism remains tricky. As with everything capitalism encounters, there is the devastating likelihood that architecture will remain subject to exploitation. Rarely do architects stand against the demands of capitalism—they would be out of work. As long as buildings continue to go up, architecture's engagement with capitalism allows for the extraction of surplus value from design services as well as the built environment. However, the fundamental creativity inherent in architecture suggests that it may be possible to disengage capital and find better ways of working. The high ideals of designers, coupled with an ethic of engagement and service could challenge the conventions of capitalist production. In any case, the intent is not to diminish the phantasmagoria of contemporary life, but to transform it into an environment that encourages the full development of all. If so, architecture would stand to flourish, as would the lives of its inhabitants.

NOTES

1   Karl Marx, 'Economic and Philosophic Manuscripts of 1844', in *Marx-Engels Reader*, ed. R. Tucker (Norton, 1978), 71–72.

2   It is interesting to note that architects are still to some degree associated by name with architecture, whereas factory workers have no association with the products they produce.

3   Karl Marx, *Capital, Volume One* (Penguin, 1992 [1867]), 322–323.

4   There are many other examples of residential developers employing well-known architects, including Charles Gwathmey, Richard Meier, Herzog and de Meuron, and Frank Gehry, who have all completed recent projects in New York City.

5   Neil Smith, *The New Urban Frontier: Gentrification and the Revanchist City* (New York: Routledge, 1996).

6   Of course these policies are to a large degree manipulated by the bond rating agencies that play a crucial role in the mechanism of capitalist exploitation. See *The Neoliberal City* (2006) by Jason Hackworth.

7   David Harvey, *Paris, Capital of Modernity* (Routledge, 2005).

8   Adrian Forty, *Objects of Desire* (Thames and Hudson, 1986).

9   Witold Rybczynski, *Home: A Short History of an Idea* (Viking, 1986).

10   Take *Dwell Magazine* for example.

11   Rob Imrie. 2004. "Disability, Embodiment and the Meaning of the Home." *Housing Studies*, Vol. 19, No. 5, 745–763.

12   Juhani Pallasmaa, "Toward an Architecture of Humility: On the Value of Experience," in *Judging Architectural Value*, ed. W. Saunders (University of Minnesota Press, 2007), 96–103.

13   Cliff Moser, "Using Active Value Engineering for Quality Management" The American Institute of Architects, (2009).

14   However, I would argue that this view treats architecture as an end in itself, which seems problematic.

15   Walter Benjamin, "Paris, Capital of the Nineteenth Century," in *Reflections: Essays, Aphorisms, Autobiographical Writings*, ed. P. Demetz (Schocken, 1986).

16   Marshall Berman, *All that is Solid Melts into Air* (Penguin, 1988).

17   Anthony Ward, "The Suppression of the Social in Design," in *Reconstructing Architecture: Critical Discourses and Social Practices,* ed. Dutton and Mann (University of Minnesota Press, 1996).

18   I am thinking of projects by Teddy Cruz, or City Farm in Chicago, as well as the many projects completed by Community Development Corporations.

19   Water Benjamin, "The Author as Producer," in *Reflections: Essays, Aphorisms, Autobiographical Writings*, ed. P. Demetz (Schocken, 1986).

20   To some degree it can be argued that Rem Koolhaas has adopted such a strategy, articulated by Dieter Lesage as "over-identification." See *Cultural Activism Today: The Art of Over-Identification*, edited by BAVO (Episode Publishers, 2007).

21   Jason Pearson, ed. *University-Community Design Partnerships: Innovations in Practice* (National Endowment for the Arts, 2002).

22   William Mangold, "Considering the Role of Van Alen Institute in Architectural Production" (unpublished manuscript).

23   These and many other organizations are profiled in *Spatial Agency: Other Ways of Doing Architecture*, edited by Nishat Awan, Tatjana Schneider, and Jeremy Till (Routledge, 2011).

3

Faster Better Cheaper: Social Process for a Modular Future

Margarette Leite

INRODUCTION

For many architects—and aspiring architects—the current economic crisis is an opportunity to redefine themselves as agents of change, improving conditions for the 90 percent of the world's population that never interacts with architects. How can we as educators inspire and support such aspirations? How can we build an academic culture with a sense of urgency and appreciation for what design can do to improve the everyday environments in which the majority of the world's population must grow and thrive? How do we connect with marketplaces that have the greatest potential to affect larger numbers of people, like the modular building industry?

Students and faculty at Portland State University are taking on the ubiquitous modular classroom in an effort to explore the value architects and design thinking can bring to the for-profit modular building industry. The major question we seek to answer is how architecture and industry can work together for the greater good.

ARCHITECTURE FOR THE GREATER GOOD

While social movements in architecture wax and wane, there is once again a clear and compelling movement in practice and education today towards a more socially responsive architecture. Spurred by recent social and economic pressures, books like *Expanding Architecture*[1] promote the works of architects and educators who use design as a tool to empower physically, socially, and politically underserved communities. Basic Initiative, Design Corps, The Rural Studio, and many university-based public-interest design programs and community design centers leverage student talent and labor while working with low-income sector communities to build housing and community service centers. Other organizations like Architecture for Humanity work directly in the public realm to galvanize public awareness and support for international causes like disaster relief. Using the global reach of the Internet, they sponsor design competitions that reach large numbers of the public.

3.1   The SAGE
Classroom, by
Portland State
University and
Blazer Industries

The work of many of the individuals and organizations touted clearly predates the current economic crisis but their relevance is made increasingly apparent as a result of it. Much of their work is supported through government funding programs, some through public and private partnerships, and many focus, rightly so, on the betterment of specific communities. The products of some of these endeavors make their way to the private marketplace in the form of the modular building industry as items to be mass-produced. But few have been able to make significant inroads in an industry so driven by market forces to produce ever-cheaper products, in ever-shorter time frames commanded by current models of consumption. And yet, mass-production can be a powerful avenue for reaching the masses of humanity that we as a profession need to address.

HOUSING AND THE MODULAR BUILDING INDUSTRY

Most people associate modular mass-produced construction with housing. Modular design and construction in the United States has progressed in fits and starts. Its earliest iterations in the U.S. were packaged houses and churches shipped from England to the colonies. Later, Sears Roebuck attained significant success with its kit house, selling 70,000 homes to growing suburban communities of middle-income America. Only with the early modern movement, however, do we find architects and critics forwarding a compelling agenda equating mass production of housing—and its focus on efficiency and economy—with social reform. With European figures like Corbusier leading the way, architects in America including Buckminster Fuller, Gropius and Frank Lloyd Wright experimented with mass-production in housing in service of progressive and utopian ideals. Although there are many companies that provide prefab, kit, and modular housing, it is the custom wood light- frame construction industry that has dominated the housing market in the US. For many reasons, the versatility of this model remains popular in spite of our awareness of the efficiencies mass-production can provide. We continue to fall behind our European counterparts when it comes to embracing modular and mass-produced components that offer quality while serving a wider audience in response to societal need. Case-in-point: the extremely popular Ikea *BoKlok* house is available as a kit or modular unit. It is targeted at and priced for single parent households and fixed-income families in Sweden and neighboring countries. In fact, prefab and modular housing makes up 90 percent of all new housing in Scandinavian countries.[2]

In the US, mass-production is, for the most part, reserved for the lowest common denominator in home provision. Manufactured mobile and modular homes are the bottom of the barrel when it comes to quality, design and comfort, so much so that to live in one is to be stigmatized. And yet mobile homes are a major provider of low-income housing in the US. In order to supply the lowest income sectors and maintain profit margins they are subject to less stringent building code requirements and are deployed with substandard materials and poor quality construction. Perhaps surprisingly then, it is in the higher end market that modular construction has more recently gained traction. Boutique houses by Marmol Radziner and the Anderson brothers stand out as exceptional examples of architects working to fully embrace modular construction. While they may find efficiencies in their methods, their clients are clearly not those of modest means. Even the Michelle Kaufman houses (now Blu Homes), so well-promoted by Dwell magazine, are targeted at middle to upper middle income brackets and remain unaffordable for most Americans.

DISASTER RELIEF AND THE MODULAR BUILDING INDUSTRY

It is in the provision of housing for disaster relief where the need for efficiency, mass production, and expediency converge. Each new tragedy mobilizes designers to produce innovative responses. Every crisis sees a number of solutions emerge and built, but generally in disappointingly small numbers incongruent with the magnitude of need. By and large the answer to housing those left homeless continues to be the infamous FEMA trailer, known for its toxicity and for its failure to address any options for permanency. Nevertheless some notable examples of alternative solutions have found limited success in the marketplace. The home improvement giant Lowes now offers a kit of parts for building the competition winning *Katrina Cottage*.

MODULAR CLASSROOMS AND THE PUBLIC SCHOOL SYSTEM

Although the plight of modular classrooms (often referred to as "portables") may not generate the immediate call-to-arms that disaster relief does, it speaks to an emerging crisis of equal scale with respect to education and childhood health in the US.

The vast majority of modular classrooms in the US are unhealthy, uninspiring and unsustainable. Nevertheless, they are the spaces in which a large percentage of our students are schooled—six million in fact.[3] In Portland, Oregon alone, portable classrooms in-use number in the thousands. Many of them are 60–80 years of age; so much for the "temporary" in "temporary classroom." They are de facto permanent solutions to overcrowding created by fluctuations in enrollment and our collective failure to provide reliable and reasonable funding to our public schools. While these problems have their roots in earlier political and economic policy choices, given our present economic and social conditions, they are likely to continue into the future.

Portland Schools

The lack of adequate funding is clearly at the root of much of what ails American public schools and Portland serves as a poster child for what has happening across the US. The history of new school construction in Portland has followed the typical cycle of growth and social change in the country as a whole with the influx and growth of Portland's population after the two World Wars and their attendant economic booms. It also experienced a virtual halt in construction after approximately 1960. In the past 60 years Portlanders have passed only two bond

3.2 Typical Modular Classroom in the US

measures to support public schools. This lack of support was influenced by the California tax rebellions of the 1970's which have had an impact on the psyche of US voters and their traditional funding priorities. In addition, while the decrease in the population of school-aged children following the baby boom generation may make newer schools seem unnecessary, the general lack of school-dedicated funding has left cities like Portland with an enormous stock of rapidly aging school buildings, many of them approaching the century mark. These buildings are in dire need of basic maintenance as well as upgrades to meet current health and safety standards. Only very recently, in 2012, did Portland tax-payers pass a 482 million dollar bond which promises to rebuild 3 to 4 schools in dire need and to seismically upgrade a number of other schools. With nearly 100 schools in the district, the amount is still well short of meeting the needs of most schools. With the results of this deferred maintenance looming, school districts struggle merely to maintain the status quo much less improve or add to their facilities. School districts, faced with space shortages, turn to operation and maintenance budgets to fund the purchase of modular classrooms because they can be considered emergency or temporary structures. Most of the structures, however, will be permanent. In fact, there are a significant number of modulars in Portland schools that date to the last cycle of school building, making them as old, and in need of maintenance, as the permanent brick and mortar school buildings.

Other factors, such as the increased mobility of American families, contribute to the rise of modular classrooms. Today American families move an average of every 5 years, which makes accurately predicting school enrollments extremely difficult for school districts. This problem is compounded by cycles of school performance (high and low) that shift popularity from one neighborhood school to another. Continually shifting enrollments in schools leave some under capacity and others overenrolled with attendant space shortages. Since current policies, driven by reduced support, pair funding allocations with student body size, some schools are reluctant to restrict their enrollments preferring to add space in the form of modular structures.

While it is tempting to think that trend of the last decade of declining student age populations in Portland may make a difference, forecasts point to a current end to the decline and a gradual but significant increase for the foreseeable future making the outlook for public education precarious and the future of the modular classroom long and healthy.

UPSIDES AND DOWNSIDES OF MODULARS

It's clear that modular classrooms fulfill an important if desperate need in the US and there are various compelling and positive reasons for our continuing reliance on them. For one, fluctuating enrollments make investments in brick and mortar schools and additions to such schools questionable propositions. Better to create a fleet of temporary, portable structures that can move as the need for them shifts. In addition, the very nature of their rapid construction, procurement, and installation serves the needs of the school system schedule. Whereas on-site construction can continue for months and create disturbance to schools in progress, off-site construction of modular structures means quick purchasing and installation schedules can happen within the time frame of the summer break. Their autonomy as structures physically separated from their parent schools can also be useful for semiautonomous uses like afterschool or gardening programs. Of course, the greatest advantage would seem to be the low cost of these structures. While the above-cited benefits are significant, the cost savings are not as great as would be expected. The average cost of a modular unit and its installation in the Portland Public School District in 2009 was between $250,000 and $300,000. It could be argued that a brick and mortar addition would not be much more. In fact, the unit itself is barely half the total cost. The rest is made up largely of site work costs, with management fees, permitting, furniture, and equipment added as well.

Since site work and soft costs are unpredictable and largely unavoidable, manufacturers of modulars are pressed continually to provide a cheaper product in order to meet schools' ever growing funding shortages. To their credit, manufacturers in Oregon have offered no-cost component upgrades to schools ordering their modulars only to be turned down because, stretched to make quick decisions and with little professional support, school districts rely on what was done previously. Thus we are faced with the current lamentable situation of uninspired and unhealthy modular classrooms parked in schoolyards.

The downsides, of course, are many, well-documented and universally maligned. They are the educational equivalent of the stigmatizing mobile homes of the US or the disaster relief FEMA trailers only they are no longer reserved for the poorest sectors of the population nor for the disaster-struck—they are now the places we school all our children, rich or poor, and they are our national shame. A considerable amount of research has been focused on modular classrooms in California, and for good reason. Of the 300,000 modular classrooms in use in American schools in 2011, 90,000 of those were in California.[4] One in every three children in California is schooled in a modular classroom. Studies attribute decreased performance and increased health-related issues to poor environmental conditions inherent to current modular classroom construction.

The negative aspects attributed to portable classrooms include poorly functioning HVAC systems that are associated with respiratory illness in children. Poorly ventilated spaces often have higher levels of CO_2 and pollutants linked to respiratory illnesses and lethargy. The source of some of these pollutants is often the modular itself as many are built with cheap, low-quality materials that have high VOC contents. Other aspects of poorly functioning HVAC systems are related to decreased performance in student learning outcomes and these include noise levels that reduce auditory clarity, inconsistent and uneven distribution of air and its concurrent discomfort to those in close proximity to air outlets.

Other undesirable attributes are associated with insufficient window openings and poor building orientation, both leading to insufficient exposure to natural daylight which is known to play a significant role in student well-being and educational performance. Environmental psychologist, Judith Heerwagon,[5] an internationally recognized figure in the field of learning environments for children and a member of our research team, has written and spoken extensively about the importance of natural daylight as well as views to nature and the outdoors to the well-being, work productivity and educational performance of human beings in general and young minds and bodies in particular. These structures are clearly not part of the sustainable and healthy futures we imagine for our children.

Having said all this, there are some bright spots in the debate. With heightened awareness of weak student performance in the US and calls for the "greening" of our schools have come some interesting proposals. Recently designers have made significant strides in bringing to market beautifully designed, sustainably produced and sustainably performing alternatives to the current modular classroom. *Project FROG* and *Gen 7* have products on the market far superior to the status quo. While they are good long-term solutions, they cannot, and arguably should not, compete with current price points on portable classrooms. As such, they may make exciting additions to some fortunate schools, but they will not be able to serve in the sheer numbers necessary to address the magnitude of the problem in the US in its current political and economic state.

But is there is a middle ground? A better product, by degrees, greener, more attuned to human comfort, more aesthetically appealing, that could find immediate acceptance in the industry and in the marketplace? There is no question that competing in economic terms with the current model is difficult. But how close can we come?

PORTLAND STATE UNIVERSITY

> Our field suffers from a tendency to elevate one application of our knowledge – building design – over all others, evident in the awards programs we run, the feature stories we publish, and the studios we emphasize in school. Thomas Fischer[6]

At PSU we are using the Green Modular Classroom project as a vehicle for looking at today's societal needs through the lens of complex social, economic, and political factors. The project was started by faculty of Architecture Margarette Leite and Sergio Palleroni in an effort to help students see that the role of architects can and should be expanded to include the exploration and definition of solutions to problems affecting the greater community. In this scenario the University serves as a forum for social change and a source of research and development efforts. The experience of the student architect becomes one of education through activism. In this case, this activism began as an investigation into the basic facts, conditions and assumptions which have defined the present role modulars have come to play. Its evolution involved further exploration through various events and educational opportunities to create a working "brief" that serves as the basis for the elaboration of a complex and fruitful social process.

While this process serves as the overall armature for our research, as well as our pedagogical goal as an educational institution, our immediate challenge for this specific project is to work with both manufacturers and school districts to find real-world solutions to problems posed by modular classrooms, solutions that will realistically address and respect the challenges to which all parties involved are susceptible. Though the tangible goal is the design and construction of a prototype classroom for Oregon that can be monitored for its performance, our broader agenda is to look at the creation of healthy and productive learning environments for children. At the same time, this challenges us to examine the oft-ignored possibilities in architecture for quality mass-production that enhances the physical environment while addressing social and societal inequities.

Process

Three simultaneous studios, looking at different issues related to modulars, became potent vehicles for promoting social responsiveness in architectural education at PSU.[7] Unlike any previous studios, they required students to operate as partners in a research-based, multi-disciplinary, collaborative enterprise to develop a socially responsive yet sound business model—not a typical skill set for most architecture students. The entire process is obviously beyond the scope of a typical quarter and unfolded into a multi-year initiative that has allowed generations of students, faculty, and professionals to contribute to the research,[8] design development, funding acquisition, and creation of a full-scale prototype in partnership with local modular manufacturer Blazer Industries.

During the first year of the initiative students enrolled in the initial studios that would define the brief. They also helped organize, and participated in, a two-part symposium sponsored by PSU and AIA Portland entitled *Learning Activism*. The symposium had two goals: to promote architecture for the public good as a form of activism, illustrated with examples from invited speakers, and to deepen interest in activism by involving participants in a real problem-solving activity on an important local issue. Among the invited speakers for the first day were national and local figures in the field of public interest architecture such as John Peterson (Public Architecture), Danny Wick (Rural Studio) and Sergio Palleroni (Basic Initiative).

On the second day, participants were given an opportunity to become activists on an important local issue. The topic was *Re-Imagining the Portable Classroom* and it took the form of an all-day charrette. Participants were presented with the issues at the heart of the debate on portable classrooms in Portland, and then worked in groups to develop ideas, sketches, and suggestions for their improvement. Participants included representatives from two leading modular manufacturers from the area, school district administrators, teachers, behavioral psychologists, architects and engineers. PSU and UO (University of Oregon) students acted as moderators for the speaker sessions and as resources and record-keepers for the charrette groups.

The symposium acted as a focused exchange that helped participating students broaden their research agendas. It also served as the basis for the initial brief that would define the continuing research by students who went on to participate in the design studio exercises that followed. In addition to participation in the symposium, students visited and interviewed students and teachers in modulars in area public schools, conducted surveys and observed the use of these classrooms. They met with behavioral psychologists to understand how children, in particular, are affected by natural light, sound, views and access to nature, concerns that are currently getting more press through the biophoilia[9] movement. They spoke with educators about current thought on new curricular models, such as *The School of One*.[10] Students toured manufacturing facilities for portables and were required to look at them from the standpoint of design, economics, construction and transportation efficiency. This experience served to make students more aware of the conditions and pressures under which manufacturers operate and which play an important role in defining the product that gets to the market.

Explorations

The design studios were divided into groups addressing different but related questions with respect to portables. In general, most of the solutions proposed could be readily implemented given the proper economic and industry incentives. Other solutions represented innovative directions that would require greater up-front research and development for longer-term efficiency but with potentially greater return. They were conducted with continued participation from many of the representatives from industry and the school districts as well as other professionals from the symposium.

One group of students looked specifically at infrastructure opportunities, exploring ways in which permanent infrastructures associated with modulars could be shared or grouped efficiently and potentially serve as platforms for a variety of uses as the need for modulars increased or decreased. Others chose to find ways to make "portables" more *portable,* to increase their usefulness to school districts that choose to address enrollment fluctuations through relocation of units. Still others chose to capitalize on the high cost of infrastructure and installation by designing prefabricated buildings that were more permanent and could grow and change with the needs of the school. A final group looked at ideas for modifying or retrofitting existing portables to make them more energy efficient and sustainably performing. They created a DIY catalogue for school districts with ideas for green wall additions, ventilation strategies, and so forth.

OREGON SOLUTIONS

These explorations in the public realm brought our concerns to the attention of Oregon's Governor's office. As an administration with a clear and well-known agenda to revitalize and green its "K-20" education, our mission appealed to them as one worthy of support as a state-wide initiative. The Green Modular Classroom project was subsequently designated an official "Oregon Solutions" project in July of 2010 by Governor John Kitzhaber. Oregon Solutions is a unique program that promotes "…sustainable solutions to community-based problems that support

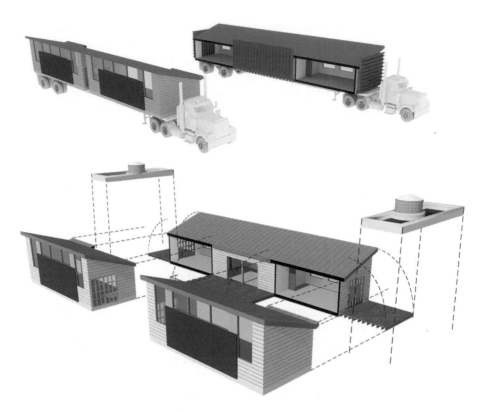

3.3   *Upgrade*, A. Green, C. Bardawil and L. Churchill

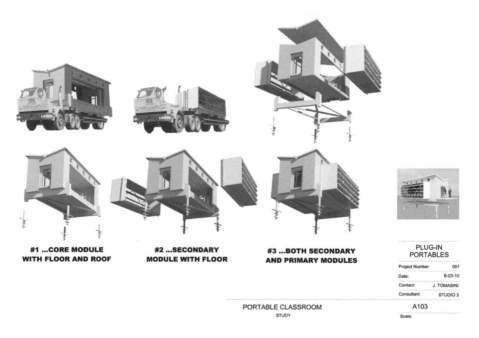

| #1 ...CORE MODULE WITH FLOOR AND ROOF | #2 ...SECONDARY MODULE WITH FLOOR | #3 ...BOTH SECONDARY AND PRIMARY MODULES |

PLUG-IN PORTABLES

Project Number:	001
Date:	6-03-10
Contact:	J. TOMASINI
Consultant:	STUDIO 3

PORTABLE CLASSROOM
STUDY

A103

Scale:

3.4   *Dropbox*, J. Tomasini

economic, environmental, and community objectives and are built through the collaborative efforts of businesses, government, and non-profit organizations."[11]

This designation allowed us to further promote and support our social and political process to address the problem of the modular classroom by expanding the community of actively involved members to include state and local building code officials, representatives from various industries including modular manufacturers, non-profit energy, sustainability and LEED consultants, Portland developers, engineers and architects in addition to students and faculty from PSU on an ongoing basis.[12]

Through this process we worked together to refine our inquiry to inform the design of a prototype solution. In doing so, we defined areas for potential savings that could help offset costs of upgrades in materials and systems as well as defining other, overall important goals associated with sustainability, performance and design for student comfort. We will refer to these as our strategies for innovation. The close working relationship that was forged through the Symposium and the Oregon Solutions process has allowed us to work closely with the manufacturers to identify opportunities for cost saving and innovation. We began by identifying current efficiencies within the system that would be advantageous to retain.

Design Strategies

Current construction practices for modular structures are maximized for efficiency in terms of materials, time, and money. They are models of lean production and finding savings in their construction costs was clearly a challenge for our team. These structures are built to maximized transportable sizes and for minimized on-site construction finishing and detailing which suits the quick installation time required by school schedules. In the current market the most commonly ordered configuration is a two-classroom unit. This configuration is provided in two long halves brought on two transport vehicles. Each half is approximately 14 ft. × 64 ft. in dimension and the halves are connected on site along a "marriage line" or "mate line" creating two classrooms of approximately 28 ft. × 32 ft. Despite its spatial limitations, this configuration accommodates a fairly large number of students per classroom. In Oregon, most are ordered as "dry" units without toilets or water facilities. We chose to accept these basic dimensional constraints and efficiencies.

In our search to find innovation we looked at the potential for shifting the costs of extensive infrastructure into creating a more valuable unit. Our strategies for innovation included:

- *Increasing Portability*   Thinking sustainably means thinking of modular classrooms as long-term investments rather than disposable structures. But that doesn't mean they need to be permanent. Improving the structure and durability of the units can make them both sturdier and easier to move. Increased portability can make them more useful in increasingly unpredictable enrollment environments allowing school districts to move them around as needed. While it is true that most modulars don't move, there are a number of reasons for this. In Portland, in particular, seismic requirements are considerable, making continuous concrete stem wall foundations that raise the unit approximately 30 inches above grade the most expedient for modular classrooms. The cost of all this concrete, the ramps required to meet that higher floor height as well as the considerable costs associated with regulatory requirements for site upgrades, permitting and project management adds up to half or more of the entire cost of adding a classroom. Moving the unit would mean abandoning a considerable amount of investment only to reinvest those same non-recoverable costs in another location. That doesn't take into account the cost associated with demolishing the considerable foundations and ramping of the original. Cost comparisons for modulars in Portland showed that surprisingly little to no savings were gleaned from the installation of previously used modulars because the lower cost of the unit was offset by higher management costs associated with the move. In general, the typical modular is not built to be moved more than once or twice. Modulars that are easier and less costly to move, and that are better built, can become more valuable commodities for sale and reuse by school districts.

- *Reducing site infrastructure*   The use of steel rather than wooden floor structures allows the modular unit both to last longer and to be located closer to the ground. This reduces the need for long unsightly ramps associated with most modulars. Shorter ramps can be made to be reused and can move with the unit. Steel floor structures can also reduce the amount of concrete required in foundation structures, reducing not only infrastructure costs, but also site impact and eventual demolition waste. Even more promising are cost comparisons showing that the use of helical pier foundation systems, while higher in initial costs, would generate significant longer term cost savings if the modular were to be moved. These foundation systems are reusable as well, which would allow them to be moved with the unit. They have virtually no impact to the site and no demolition waste when later removed. The inclusion of reusable ramps and foundations as a part of the unit rather than as permanent site infrastructure, again contributes to the creation of a commodity of greater value for resale and reuse by school districts while also reducing the burden on the environment.
- *Rethinking HVAC*   Most HVAC systems in modular classrooms are sized in order to provide the amount of ventilation required by code to properly supply a fairly large number of students in a relatively small room. This often makes the systems oversized for heating and cooling purposes contributing to noise and air disturbance issues that are common complaints with modulars. Our team took a "whole building design" approach to systems design by considering the students in the classroom as a primary thermal resource. It doesn't take much time for numerous active bodies to generate enough heat to provide the necessary heating. As the needs of the classroom are driven primarily by fresh air provision, this can be provided simply by Energy Recovery Ventilation (ERV). The bonus effect is that ERV's can provide

considerably better quality of air, and more air exchanges than tradition HVAC systems. In addition, in a temperate climate like that of Portland, air-conditioning is generally an unnecessary and energy consumptive exercise, particularly during the school year cycle when most modular classrooms are in use. Extensive energy modeling by our students in conjunction with local engineering firms showed that we can provide an indoor environment that meets the comfort zone throughout the colder months of the year without the traditional heat pump unit. Instead we propose using only the ERV along with a heating coil to kick start heat in the morning. For cooling, natural ventilation is induced by placement of operable windows, ceiling fans, and the use of a phase change material made of a wax-based substance that is placed within the stud wall cavity and is the equivalent of 4 inches of concrete in its ability to slow heat transfer through the walls. This system serves as our base model. In cases where summer use is critical or where climatic conditions are different, higher efficiency heat pumps can reduce energy usage and noise disturbance. As an additional upgrade, ventilation can be boosted with roof mounted, solar powered venting.

- *Celebrating Natural Light*   In Portland, where sunny days are a cause for celebration, particularly during the school year term, natural light is a resource to be gathered wherever possible. The inclusion of a clerestory along one long side of the modular classroom, along with well-placed, larger view windows can nearly eliminate the need for artificial lighting (which also reduces energy consumption) even on the grayer days of the year. Proper orientation, however, is also critical. Because the typical modular classroom is designed without a specific site in mind, it is orientation neutral and therefore operates independently of natural light with few windows only to provide views. It is dependant, year round, on artificial lighting. The simple inclusion of additional framing to allow for doors on either side of each classroom allows the portable the flexibility to be placed with the clerestory in the ideal orientation to provide natural lighting while allowing the entrance to be placed wherever it meets the needs of the school. It also creates the possibility of a second door for the addition of spaces like bathrooms, breakout rooms and decks at a later date when additional funding may be more available. Proper orientation also reduces solar heat gain and loss, but most importantly, it can provide a continuously and naturally lit environment along with abundant view windows in support of student well-being and performance.

In addition to these major areas for innovation, the green modular will include low to no-VOC materials, hi-efficiency fluorescent lighting with sensors and dimming controls, low-E, dual pane windows with select operables, cork wall surfaces for pinup and acoustical control, as well as exposed ductwork.

As previously stated, the expected outcome of this process was the production of a full-scale prototype. While we have worked with local school districts to discuss design concerns and issues, this project was undertaken as a speculative

3.5   The SAGE
Classroom, by
Portland State
University and
Blazer Industries

venture without an identified "client" but with a purpose to transform the status quo by example. This has allowed us the distinct advantage of doing a clearly evidenced-based research and design process, a unique opportunity among public/ private initiatives of this scale. The prototype was invited to be displayed at the 2012 Greenbuild Conference in San Francisco. The unit, now dubbed the SAGE classroom, for "Smart Academic Green Environment," was purchased by Pacific Mobile Structures Inc., which will distribute the classroom throughout the Northwest US as well as British Columbia and Alberta. In addition, the unit is contracted with two other distributors across the country, Satellite Shelters Inc., across the midwest and southern US, and Triumph Modular in the northeast part of the country, with the first classrooms rolling out to a number of schools in early 2014. As for our work with the industry, while it might be natural to expect some hesitancy on the part of industry to work with architects, particularly with student architects with little experience, we have, in truth, encountered a great deal of support. Manufacturers are aware of how their products are perceived by the general public and are sincerely interested in working toward improving them, but their constraints and priorities are not always obvious to architects. As in any collaborative process, the relationship must be one of mutual respect and exchange of knowledge. The involvement of students may in fact eliminate the potential perception of an imposed agenda that inevitably accompanies outsider involvement in another's territory. In this case it is clear that the students are learning a great deal from industry participation and are not reticent to admit it. However, concerns about working with the private for-profit sector on our part do exist. The question must be raised: as an academic institution concerned with providing an educational opportunity rather than seeking remuneration, are we simply providing free labor and design talent to companies for their own financial gain?

CONCLUSION

The Green modular classroom project has served to reveal that the following:

- That architects can and should step up as leaders in tackling larger social issues that may be caused or aggravated by poorly addressed physical environments.
- That public and private interests can come together through a socially driven process to address common social concerns.
- That architects can work in a collaborative and productive way with the modular building industry to address the needs of larger populations for affordable but healthier and more sustainable solutions.

It is critical today that architects begin to see themselves as potential agents of change in this world of shrinking resources, expanding needs and great uncertainty. It is a chance to expand our perception of architecture to include the mundane environments that are the "architectures" of the majority of people around the world and to find ways to serve broader segments of the population. While it is seems obvious to look to mass-production as a means, it has also been proven throughout history to have little impact in the US as a tool for creating social change or social equity. In Europe, prefab and modular construction seems to have captured some of the spirit of architects of the modernist, utopian movements. Here, for the most part, modular and mass-produced construction seems motivated only by the goals of efficiency, expediency and profit without the loftier goals of social reform and social justice to ground it. And yet this hope on the part of architects continues to resurface in the literature and through the projects of crafty architects whose designs are exciting yet still largely unattainable. To propose once again that prefabrication and modular construction is about to take off in this country is generally laughable[13] and yet… could the current economic and social climate be a turning point? Will our new collective obsession for "green" architecture be the final impetus? We know that offsite prefabrication can reduce construction waste as well as reduce impact to the environment. Indeed the use of prefabricated parts and whole modules can help score LEED points. There are an unprecedented number of projects now touting their "green-ness" through the use of modularized and prefabricated systems. The National Research Council recently created a committee to identify its goals for improving the efficiency and productivity of the construction industry in the United States. It identified 'Greater use of prefabrication, preassembly, modularization, and off-site fabrication techniques and processes'[14] as a major breakthrough area. The US Modular Building Institute (MBI) rather conveniently forecasts for itself a 5% increase in modular construction in the US in the next two years alone – not huge but in the current real estate climate fairly optimistic.[15]

For all our fascination with modular construction, why is it so little taught in our schools of architecture? If we don't begin to inform ourselves in the use of more efficient systems of architectural delivery, we risk finding no place for ourselves in an industry that seems happy to proceed without our input and without the human based, socially-focused imperatives that we can bring. Can we use a more socially-centered process like the one described here to reintroduce human and evidence-based design research to the profession and in a new partnership with industry? We need to begin by educating students to see the value in these kinds of processes.

Lastly, it is important to note that this process serves to express to the next generation of architects that the enticing, formalistic exercises that still make up the majority of design education in this country do not represent the whole range of possibilities for practice in the field nor do they give voice to the more pressing concerns we should be addressing as designers. The good news is that students of today are keenly aware of the dire situation we as a global society find ourselves in with respect to the environment and to issues of social justice, and are not the hard sell we envision them to be.

BIBLIOGRAPHY

Anderson, Mark, *Prefab Prototypes: Site Specific Design for Off-Site* Construction (Princeton Architectural Press, 2007).

Annual Report on Relocatable Buildings, Modular Building Institute, 2011.

Annual Report on Permanent Modular Construction, Modular Building Institute, 2011.

Architecture for Humanity, Open Architecture Challenge: Classroom, 2009.

Bell, Bryan and Wakeford, Katie, *Expanding Architecture: Design as Activism* (Metropolis Books, 2008).

Coates, Ta-Nehsi, 'The Littlest Schoolhouse', Atlantic Monthly, Jul/Aug (2010).

Fischer, Thomas, 'The Once and Future Profession', *Archvoices,* AIA Online (2002).

Fulcher, Merlin, 'Prefab Schools Debate Heats Up', *Architects Journal*, October (2010).

Hawthorne, Christopher, 'Prefab: the Dream that Refused to Die', Metropolis (2011).

Heerwagen, Judith, 'Psychosocial Value of Space', Whole Building Design Guide (National Institute of Building Science, 2008).

Johnson, Kirk, 'School District Bets Future on Real Estate', *New York Times* (Sept. 4, 2012).

Modular Building Institute White Pages.

Oregon Solutions website, http://orsolutions.org/.

Wilson, E.O., *Biophilia* (Harvard University Press, 1984).

NOTES

1   Bryan Bell and Katie Wakeford, *Expanding Architecture: Design as Activism* (Metropolis Books, 2008).

2   2011 Annual Report on Permanent Modular Construction, Modular Building Institute.

3   *Architecture for Humanity, Open Architecture Challenge: Classroom, 2009.*

4   2011 Annual Report on Relocatable Buildings, Modular Building Institute.

5   Heerwagen, Judith, 'Psychosocial Value of Space', *Whole Building Design Guide* (National Institute of Building Science, 2008).

6   *Fischer, Thomas,* 'The Once and Future Profession', *Archvoices, AIA Online,* 2002.

7   Studios taught by Margarette Leite, Sergio Palleroni and Tim Dacey.

8   Research for this project was supported by a grant from the Institute for Sustainable Solutions (ISS) at Portland State University.

9   Wilson, E.O., *Biophilia* (Harvard University Press, 1984).

10   Coates, Ta-Nehsi, 'The Littlest Schoolhouse', *Atlantic Monthly*, Jul/Aug (2010).

11   Oregon Solutions website, http://orsolutions.org/.

12   The Oregon Solutions Green Modular Classroom Team includes PSU's Schools of Architecture and Engineering, The Green Building Research Lab at PSU, The Institute for Sustainable Solutions at PSU, Blazer Industries, Pacific Mobile Structures, Inc., Gerding Edlen, AIA Portland, Portland Public Schools, Oregon State Building Codes, Portland Bureau of Development Services, Northwest Renewable Energy Resources, Energy Trust of Oregon, Oregon BEST, PAE Engineers, McKinstry Engineers, M Space Holdings, EcoReal and Luma Lighting.

13   Note articles like Christopher Hawthorne, 'Prefab: the Dream that Refused to Die', *Metropolis (*2011).

14   Modular Building Institute White Pages.

15   Modular Building Institute White Pages.

Re-defining Architectural Performance—*Survival Through Design* and the Sentient Environmentalism of Richard Neutra

Franca Trubiano

CHALLENGING PERFORMANCE METRICS

> *"If design, production and construction cannot be channeled to serve survival,*
> *if we fabricate an environment… but cannot make it an organically possible*
> *extension of ourselves, then the end of the race may well appear in sight. "*
> *Richard Neutra*, Survival Through Design

During the past two decades, the architectural profession has taken a decidedly technical turn. Architects have re-engineered their practice by introducing modes of enquiry and methods of design attendant to the fields of building physics, thermodynamics and material chemistry. The language of simulations, energy modeling, benchmarking, design assist, project delivery, integrated practice, energy conservation, life cycle costing, and building information modeling permeates the architect's daily practice in North America and Europe.

The larger conceptual framework subtending this activity is the desire to design, construct, and operate high performance buildings that consume less energy and material resources and in so doing, reduce their environmental impact. Advancement in the field relies upon one's access to a comprehensive set of metrics, measurements, and methods for quantitatively assessments of a building and its engineering systems. Using agreed to methods of analysis, with results benchmarked against industry wide standards, high performance designs are called upon to contribute to better operating buildings that moderate the global output of life threatening carbon and greenhouse gas emissions. To this end, a large number of organizations, associations and research institutions are committed to identifying said metrics, including a sub-sector of "green" related service providers dedicated to the integration of advanced energy systems.

Notwithstanding these substantial efforts, the field has yet to identify a comprehensive and operative definition of "high performance" that is acceptable to architects, engineers and builders alike;[1] practices deemed essential by one building professional, are often disparaged by another. Building sensors used by engineers to measure occupancy and CO_2, seem of little consequence to architects intent on improving the quality of natural lighting in a room. Integrating water and waste management within a building's sustainability profile may be important for elected officials, but only tangential to engineers concerned with energy reductions. For some, this variety of approaches contributes to an enrichment of the field, fostering a diversity of methodologies. For others, the inability to define an agreed to baseline for attaining high performance, shared by all professionals who design, construct, and operate buildings indicates a far more disquieting situation; this being, the field's near fetishistic deployment of techniques and tools for the amassment and manipulation of data at the exclusion of all other methods of evaluation.

More than ever before, diagnostic building descriptions are gathered using spreadsheets, equations, and algorithms. Numerically centered data is used, almost exclusively, for all types of building assessments. And notwithstanding the obvious advantages of knowing more rather than less about the energy and material consumption of a building, rarely is critical discussion brought to bear on whether the pursuit of high performance buildings contributes to a more ethical, sustainable and socially relevant practice of architecture. Implicit is the assumption that using less energy and fewer materials automatically results in a more sustainable environment. But this may not, in fact, always be the case. Few are the true opportunities to gage whether the search for "high performance building" metrics, predicated on ever expansive data-scapes, actually ameliorates the human condition. As a practitioner in the field, this remains a difficult predicament.

Addressing this issue requires a far more comprehensive definition of performance; one better suited to addressing the many challenges now faced by the built environment. What is sought is a characterization of the field that repositions the sentient being at the very center of its enquiries. While survival of the environment is an essential goal of high performance design, so too is the survival of human life. And to this end, a subtle recalibration of methods may be required in order to foreground a practice of "performance" that operates beyond building. To initiate such a shift in the discursive character of the field that is "high performance", this essay turns to the theoretical project set forth by Viennese/American architect and author Richard Neutra (1892–1970) who, with words and works, subscribed to a vision of environmental design of significant import to this discussion. An early advocate of objective and verifiable measures of design, Neutra subtended his own definition of "performance", albeit one more particularly aligned with the metrics of human biology, physiology and neurology. By way of a critique of mid century building practices, Neutra promoted a science of design performance that was inclusive of human perception and responsive to embodied sentience. This is the particular aspect of his intellectual activity that this paper addresses, with a view to challenging our contemporary understanding of building performance.

SURVIVAL THROUGH DESIGN

Neutra practiced during the second and third quarters of the 20th century, having made significant contributions to the spatial definition of modern architecture.[2]

Alongside his colleagues Rudolf Schindler, Charles and Ray Eames and Raphael Soriano, he helped define a vision of contemporary residential living, whose desired reconciliation with nature was unencumbered by the weight of tradition.[3] An early advocate of built environments designed for maximum functional ease and clarity of purpose, Neutra's homes—sited principally on the west coast of the United States—codified many of the design parameters now characteristic of the mid-century American residence.

A consummate writer, in 1954 Neutra published the first edition of *Survival through Design,* a work of architectural theory that addressed many of the most pressing questions affecting contemporary design in post war America.[4] The book was reviewed over half a dozen times in the year following its publication, but more than the number of reviews, of interest is how widely it was examined in art history, sociology, and political science journals.[5] Its thesis had attracted many far afield from architectural theory who recognized in Neutra's hypothesis an environmental orientation resonant with the humanities and social sciences. Ivan Johnson's review, for example, published in *Marriage and Family Living,* foreground the book's thesis that poorly conceived housing could have long-term effects on the nervous condition of inhabitants and pose a health risk to families. Whereas Paul Pfretzschner's review in *The Western Political Quarterly* encouraged political scientists to recognize the possible implications, which industrialization and the quality of the built environment had on the larger political order.

Survival through Design received far less consideration, however, from the architectural press of the day, not having been reviewed by architects despite extracts appearing in advance of the book's release in the January 1954 edition of *Architectural Forum.*[6] Neutra's buildings had been widely featured in over fifteen Italian, German and English journals in 1954 alone; however, the value of his theoretical speculations was of little interest to designers. Except perhaps for art and architectural historian Sibyl Moholy-Nagy, who reviewed the book in the *College Art Journal* of 1954.[7] Having, herself, but a year later published on design, environment and performance, Moholy-Nagy captured the essence of Neutra's environmental argument in recognizing the interconnectedness of human sentience and the built environment. Her review substantiated the value of his thesis proposing the human mind capable of processing a vast array of sensorial stimulants and, by a form of inductive reasoning, capable of near objective responses whose analysis and rationalization were vital to the science of design.

Neutra was determined to resituate the measure of architectural quality within an understanding of the built environment that was both responsive to nature and responsible for its continued health. Surely all forms of life were threatened by poorly designed environments, but no more so than human life itself. He theorized a direct reciprocity between human subjects and their environments, both natural and man-made. On the one hand, it was well known that the quality of natural environments contributed to the quality of the human condition; different climates, material resources, and geographical topographies shaped human sensibilities in a variety of ways, throughout various parts of the world. However, less well recognized, was mounting evidence from the scientific community that built environments had an equally determinant effect on the biological, physiological and neurological dimensions of human life. To this body of knowledge, Neutra had been highly sensitized through the work of geneticists, anthropologists and arctic explorers.[8] As such, works of architecture were never mere shelters, but rather strategic participants in a science of design; the study of whose variables were central to his definition of performance and essential for long-term survival.

4.1 and
4.2   Exterior
and landscape
view of the Alfred
de Schulthess
House, Richard
Neutra Architect,
built in 1956, in
Cubanacan Playa,
Havana, Cuba.
By permission of
the Bundesamt
für Bauten
und Logistik,
Government
of Switzerland,
Photographer
Nathaniel T.
Schlundt

SEEKING EXPANSIVE MEASURES FOR PERFORMANCE

Like many architects of the post war period, Neutra debated the existence of an objective measure of architectural value.

> *"Are there reliable values which are at least sharply silhouetted against the horizon of the future? Can we define such values beyond those which are commercially advertised? Can we make these values more soundly founded or defensible? How is the knowledge of these values to be obtained with a degree of assurance?"*[9]

The desire to establish impartial measures for architectural design had captivated the architect's imagination since the 17th century.[10] In *Survival through Design*, Neutra acknowledged, in a brief but comprehensive review of the origins of rationalism in European architectural theory, the continued absence of agreed to metrics for architectural excellence.[11] Traditional definitions of architectural "value", based on belief in innate "qualities" and, current as late as the 18th century, were of little use in a world now dedicated to reason and the verifiable truths of science.[12] Administered hierarchically and by exclusive right of privilege, architectural "qualities" codified in *ancien régime* academic theory, were inappropriate for an industrial democracy of the late 20th century.[13]

A new definition was required which foreground the social context of post-war America. Operating within a highly industrialized consumer based economy; Neutra was unsentimental about transferring the power of criticism from the privileged few to mass consumers, recognizing great promise in their ability to judge the merit of a building. Everyday users, uninitiated in the classical tradition of "qualities", evaluated the built environment in direct response to how it served their immediate needs.

> *"By bitter experience, the consumer of our civilization learned that he had no means of judging alleged qualities; he could judge performance only."*[14]

No longer would a work of architecture be judged by methods favoring the rules of decorum. Henceforth, the only true guarantor of excellence was the object's "use" and its ability to perform to a desired function. The term Neutra selected to describe this new barometer of quality was "performance", which albeit not the first ever use of the word by an architect, succeeded in repositioning architectural judgment within the realm of the lived. Challenging more than 150 years of aesthetic theory, whose main intellectual legacy had been to promote the disinterested contemplation of the arts, Neutra resituated architectural value within the domain of "use".[15]

With little nostalgia for the loss of a craft based building practice, he acknowledged the technological horizon of this new definition of value;

> *"Good old qualities could not be preserved, still less arbitrarily revived. The machine age called for the creation of a new sort of life and a new type of quality."*[16]

Modeled on transformations taking place in industries such as the manufacturing and marketing of automobiles, Neutra's language revealed a significant shift in the referential dimension of architecture.

> *"quality specifications have been replaced by performance specifications, that is, by a description of the performance capacity and operational objective."*[17]

> *"the buyer of an automobile seldom knows what is inside the engine housing, nor does he hire an expert to find it out… What is actually given him or what he asks for from the supplier's agent is a performance guarantee."*[18]

So stating, the quality of a constructed object was less determined by the values of its craftsman than by those of its manufacturer and users.[19] In fact, this definition of performance, centered on the language of "specifications" and guarantees, was not new or singular. Buckminster Fuller had a year earlier called for the design of high performance dwellings in his description of the experimental design/build initiative that was the Yale Collaborative of 1952 and in his work for the "Standard of Living Package."[20] Invited by Yale to create and innovate in the matter of housing, Fuller theorized the use of prototyping at the industrial scale to achieve his goal of delivering "high-performance" dwellings. Invited by MIT during the same year to participate in their 1952 Housing Conference, Fuller claimed to *"have tried to carry on a scientific prototyping activity to show how the house product can be designed for performance."*[21] Almost a decade earlier, in 1947, architect Theodore Larson had promoted the adoption of performance-based standards, familiar with their success during the war. Larson called for their immediate implementation in post war industrialized housing as rules of thumb were no longer adequate for meeting the disproportionate demand for shelter in a building economy increasingly committed to the use of highly engineered materials.[22]

Neutra, too, had been sensitized to large-scale transformations in the production of building materials due to industrialization and the rapid expansion of consumer driven markets. The very concept of building with raw materials was a thing of the past as was the idea of a *"unity of material"*.[23] By mid century, Neutra recognized that the majority of building materials were henceforth engineered and manufactured to performance standards such as those promoted by the American Society for Testing Materials (ASTM).[24]And, notwithstanding the extent to which building assemblies were overwhelmingly motivated by a desire for standardization, this shift in manufacturing practices was in no way tantamount to the *"vulgarization"* of architecture. Rather, the large-scale mechanization of products ensured all building technologies attained *"standard specifications, always focusing their interest on performance."*[25] After all, the capacity to produce a vast number of commodities destined for mass consumption only increased the number of opportunities to achieve what he termed *"machine-made perfection."*

> *"The refinement of the post-aristocratic quality type is completed only when commodities, tested through performance, can be produced in mass because they are sold at a nominal price."*[26]

Having chosen the example of an affordable every day light bulb, Neutra described its beauty as the result of many minds simultaneously at work in the making of a single perfected object; its design, a form of collaborative research implicating inventors, producers and consumers. The invention was borne of "*hundreds of laboratories, … and perfected through thousands of resourceful and highly trained brains.*"[27] Far from the invention of a singular mind, inspired *ex nihilo*, the inventor/designer of the most ubiquitous "objet type" was the collective mind of society and technology. More pointedly, Neutra's insistence on the communal and participatory origins of the light bulb, recalls Kevin Kelly's own description of its beginnings as a pseudo biological event of convergence. In his book, *What Technology Wants?*, this chronicler of information technologies asserts that the invention of the light bulb was a clear example of creative simultaneity in which a similar idea was born of numerous minds and disparate locations.[28] Notwithstanding differences in climate, temperament and resources, inventions—technological and otherwise—appear contemporaneously as if the result of a communal wish fulfillment.

For Neutra, simultaneity was equally constituent of "use"; the adoption of an object of design by more than a billion individuals surely attributed to it a measure of excellence, which he termed its "standard performance".

> "*It is the potential and the actual market of a-billion-a-day users of electric bulbs that brought into existence this new type of quality, a quality difficult to understand in itself, but easily appreciated in a standard performance.*"[29]

An entirely new premise within which to interpret the role of "use" in architectural design had been defined. Associated with the concept of "performance", this characterization of "use" transcended both its classical origins in the Vitruvian term "*commoditas*", describing the appropriateness of a building's plan and organization to the activities it housed, and its more modern affiliation with "function", the term with which it was most often confused in the early 20th century.[30]

Central to Neutra's definition of standardization was the value an object accrued when useful to the greatest number of people, for "*only users could assess true value.*"[31] The purchasing power of billions had ushered a new definition of excellence. And while it would take five more decades before the pursuit of standardization was transformed into a desire for customization, the potential impact, of a massive increase in building "users" on the measure of architectural value, was clearly recognized by Neutra in 1954.

Neutra was, however, no less captivated by the changing characterization of the "user", from the individual to the many. Owners, occupants, and clients were no longer private individuals committed to the design of a one-off commission, but more typically representatives of a group, a company, a congregation, or an institution. Increasingly, architects were called upon to conceive of projects for larger constituencies, including;

> "*… city fathers who dispose of staggering tax revenues extracted from all of us,*" and "*huge industries, … politicians and bureaucrats who pass on a big, mixed bundle of designs for … city halls, schools, projects for hundreds of thousands of families, and on plans to develop entire regions, states or nations.*"[32]

The purchasing power of post-war US families, government bureaucrats, and corporate technocrats had replaced the reign of the exclusive and the rare. Projects with significantly larger scopes had become commonplace during Neutra's career, particularly in the State of California whose census records first noted a population of over 90,000 in 1850; 3.4 million during the 1920s when Neutra first immigrated to the US; and 10 million when the architect was building and publishing in the 1950s.[33] In this regard, one need only recall his description of the challenge he faced when called upon to design for a *"new scale of faith"* that was the Community Church in Orange County California, *"an expression of our contemporary world-culture."*[34]

Fulfilling the needs of such sizeable numbers necessitated new theories and techniques. And to this end, Neutra advocated the adoption of *"probability and the statistical point of view,"* means which would actualize the needs of so many.[35] For only in this way could the vast quantity of information associated with so large a group be brought to bear on the design process. The "gigantic" scale of building necessitated alternative methods whose scientific metrics facilitated the *"…wholesale planning 'of large bodies', such as communities, or regions."*[36] Henceforth, design in the form of trial and error would no longer suffice for capturing the sheer speed and scale of modern life. A re-calibrated practice of design required *"more precise and pertinent data"*[37] because *"staggering masses"* live at *"such high velocities."*[38] And, in this utterly modern context, architects would deploy near scientific hypotheses, *"recommending certain paths of research that lead to objectively verifiable results."* So doing, architects could *"prove anything to councilmen, taxpayers, administrators, boards, or the people."*[39] Neutra had associated architectural excellence with the aggregation and manipulation of data, promoting the veracity of a performance standard in the input of large users.

For many, Neutra's position appears fully consonant with our own culture of numerically based proofs for validating buildings; a practice fully predicated on verification, simulation and statistical analysis. And yet, Neutra's thesis was far different, in one fundamental way. His decidedly technical turn had defined the language of performance using measures far beyond the domain of buildings. Neutra repositioned the faculty of judgment within the most elemental definition of the human condition. True measures of performance resided not in the physical givens of a constructed artifact, but in the far-reaching capacity of sentient beings to respond to the built environment in which they reside. The ability for mind and body to interpret the full range of sensorial stimuli emanating from a given environment defined the science of human perception. And it was critical to his argument, that objectivity in the projection of forms and in the study of their functions was only achievable when *"defensible values obtained through a degree of assurance"*,[40] were aligned with the sciences of biology, physiology and neurology. For Neutra, performance standards transcended the purely technological to engage, more substantially, the demands of human physiology.

> *"… data will have to concern above all proved and clarified common human potentials. If our designs are to hold water, we not only must have a technological and commercial horizon but we must truly know man, the consumer, and his 'physiological purchasing power.' To plan for him we must know his characteristics."*[41]

In this context, the collection of data and its incumbent numerical proofs were activities less concerned with the operation of buildings than with the capacity for design to sustain human life.

MEASURING BIOLOGICAL, PHYSIOLOGICAL AND NEUROLOGICAL PERFORMANCE

Surely, it is our destiny to make tools and to build. However, as Neutra noted, our environmental predicament has only been hastened by continued neglect of incontrovertible evidence of the dual character of survival. While on the one hand, humans undermine the natural balance of eco-systems with their constructions; on the other, built environments have a direct and lasting affect on the human condition. No substantial divide exists between nature and artifice. In holding to the tenet we are separate from our environment, human organisms continue to be threatened by their own actions. After all, have we not with our tools fashioned a veritable new nature? According to Neutra, *"in recent times the specifically human capacity of troubling nature has increased beyond all the artificialities of the ancient régime;"* this, clearly evident in the power of alpha and gamma rays *"to influence even the most chromosomatic base of the species and cause heretofore unheard-of mutations."*[42] Hence, by the artifice of science, as by the artifice of design, we've developed *"an entirely new environment"* which threatens the very essence of human biology, physiology, and neurology.

> *"Conceivably far-reaching influences on the future of a species can be exerted through design. Out of ignorance, we permit our instrument, human design, to operate accidentally, and it may bring about mutations more faithful than nature's."*[43]

For Neutra, insuring *Survival through Design*, required acknowledging the degree to which built environments had already contributed, *"out of ignorance"*, to mutations harmful to human life. By 1954, 200 carcinogens were known to the author - all by-products of invention, technology and design—and disconcerting was the capacity for sensory adaptation to even the most deadly of poisons. In a statement foretelling of Al Gore and Davis Guggenheim's cartoon vignette in, *An Inconvenient Truth* (2006), which portrays a frog who, when placed in a pot of hot water, learns to adapt to increasing temperatures unaware of the eminent danger of boiling, Neutra argued our very survival was at stake if we remained uninformed of the science of sensory response.

> *"For survival, we cannot always depend on our senses. They often fail to report danger in the smallest dose, which sometimes is the most dangerous. … The fact that a man does not realize the harmfulness of a product or a design-element in his surroundings, does not mean that it is harmless. We need other, more objective, criteria than mere opinions or custom and habit."*[44]

In this statement that predates by nearly a decade Rachel Carson's seminal work of environmental theory—*Silent Spring*, Neutra unknowingly characterized himself as an early environmentalist. His concern was not however with the effect of pesticides on the environment, but rather with the effect that badly conceived buildings had on human physiology, biology and sentience.

> *"The man-made setting reacts through an infinite number of stimuli upon the nervous system of every member of the community. More than that, today design may exert a far-reaching influence on the nervous make-up of generations."*[45]

Speed, the wasteful consumption of energy and the displacement of large quantities of matter have *"created a biological situation without precedent."*[46] Our transformation of the environment has been so extensive and so categorical, the body has barely had an opportunity to react and respond. And because human adaptation to external man-made artifacts is a far more complicated process to decipher, decode and rationalize than *"biological adjustment to a natural habitat,"* more objective measures were required to fully comprehend the ways in which human biology, physiology, and neurology were directly affected by the built environment.[47]

According to Neutra, when designs are constructed they are capable of generating pleasure as well as inflicting pain. The capacity of the brain's neuro-receptors to decipher one from the other is essential for the practice of design, as it is for the survival of humankind.

> *"Any design that impairs and imposes excessive strain on the natural human equipment should be eliminated, or modified in accordance with the requirements of our nervous and, more generally, our total physiological functioning."*[48]

The ability of our physical and mental faculties to react positively to the built environment is an all-important principle to be upheld when evaluating the merit of man-made artifacts. For similarly to the way in which food and drugs are carefully regulated for consumption, so too should the contemporary consumer be safeguarded from the harmful effects of poorly designed buildings.[49] To this end, Neutra was intent on attributing objective measures to the practice of design, and in only one example, he identified three possible types of stimuli one encountered when confronting a man-made surface. The first was the kind of stimuli which resulted from the experience of a "continuous" and uniform pattern; the second, a more syncopated stimulus made possible by a surface with "rhythmical" characteristics; and the third, an "irregular" form of response, said to be rarely preferred over the second and the first.[50]

Articulating such an analytical framework for the sentient response of human beings to the built environment, only demonstrates the extent to which Neutra believed our faculties of perception truly quantifiable and verifiable. He encouraged the outright scientific testing of the body's *"sense receptors"*. Not only the eye, but the *"total nervous system"* could be understood vis a vis the environment to which it was subjected.[51] Only a fully kinesthetic understanding of perception, which foreground a building's contribution to *"general tenseness or lack of relaxation, …increased or reduced receptivity to additional stimuli, …glandular and digestive secretion, …[or] modified metabolism,"* could alert us to the dangers of survival.[52] And therein, Neutra proceeded to describe the relationship which moisture, air movement, heat, materials, body pressure, gravity, and odors all had on human perception and architectural quality.

> *"Organically oriented design would, we hope, combat the chance character of the surrounding scene. Physiology must direct and check the technical advance in constructed environments."*[53]

RE-ENGINEERING THE METRICS OF PERFORMANCE

For Neutra, a detailed examination of the human sensory apparatus was absolutely necessary for defining a veritable science of design. Rigorous and concentrated efforts were needed to identify an objective analytical language whose parameters, predicated on the kinesthetic dimension of the senses, could be employed to design works of architecture fully consonant with life. Comprehending the body's biological, physiological and neurological capacity to assimilate stimuli from the built environment, both positive and negative, was a factor of dire consequence for the survival of the species.

Having already defined the human subject at the scale of mega measures when championing the value of a billion "users" for defining performance standards, his attention to the science of sentience, contributed yet another scale—albeit this time, that of micro measures. In a manner prescient of Charles and Ray Eames' *Power of Ten* (1977), Neutra had articulated a framework for architectural thinking bounded by dimensions exponentially large and exponentially small; reconciling the desire for human survival at scales simultaneously celestial and physiological, global and neurological.[54]

His promotion of a human centered definition of performance, whose metrics were simultaneously aligned with the needs and desires of the many and the one, of the macroscopic and the microscopic, was the particular insight of Neutra's position. In a slight reconfiguration of the givens, he reminded readers that the pursuit of objective measures was of most value when in service to the survival of all species. This lesson continues to hold much currency for our own practice of high performance design. After all, of what "use" are high performance buildings devoid of any and all life?

NOTES

1   For a larger discussion on the definition of 'high performance', see Franca Trubiano, "High Performance Homes: Metrics, Ethics and Design", in *Design and Construction of High Performance Homes – Building Envelopes, Renewable Energies and Integrated Practice* (Routledge Press, 2012); pp. 3–22.

2   See Thomas Hines, *Richard Neutra and the Search for Modern Architecture* (Rizzoli: New York, 2005), David Leatherbarrow, *Uncommon Ground, Architecture, Technology and Topography* (MIT Press, 2000), Sylvia Lavin, *Form Follows Libido, Architecture and Richard Neutra in a Psychoanalytic Culture* (MIT Press, 2005).

3   See Wolfgang Wagener, *Raphael Soriano* (Phaidon, 2002) and Judith Sheine, *R.M.Schindler* (Phaidon, 2001)

4   Richard Neutra, *Survival through Design* (Oxford University Press, 1954–1969).

5   See Charles A. Ascher, "Survival through Design by Richard Neutra" *Academy of Political and Social Sciences* (September, 1954), pp. 182–183; Dorothy Grafley, "Survival through Design", Weathervane section, *American Artist* (November 1954), p. 48, 56–58; Contance M. Perkins, "Survival through Design by Richard Neutra", *Journal of Aesthetics and Art Criticism*, Vol. 13 (1954), pp. 273–74; Paul A. Pfretzschner "Survival through Design by Richard Neutra", *The Western Political Quarterly*, Vol. 8 (March 1955), pp. 146–147; Johnson Ivan, "Survival through Design by Richard Neutra", *Marriage and Family Living*, Vol. 17. (1955), pp. 181–182.

6   "Excerpts from Survival through Design, by Richard Neutra," *Architectural Forum*, Vol.100, (Jan 1954), pp. 130–133; "Survival through Design", *USA Tomorrow*, Vol. 1, (Oct 1954), pp. 74–79.

7   Sibyl Moholy Nagy, "Survival through Design by Richard Neutra", *College Art Journal*, Vol. 13. n. 4 (Summer 1954), pp. 329–331. For Moholy Nagy's discussion on the subject of performance, see Sibyl Moholy Nagy, "Environment and Anonymous Architecture", *Perspecta*, Vol. 3 (1955), pp. 2–7, 77.

8   Neutra cited the work of a number of scientists including Dr. Tracy Sonneborn at the University of Indiana (p. 83) who studied biology and genetics, Russian anthropologist and ethnographer Vladimir Jochelson (p. 89) and Arctic explorer, Dr. George Murray Levick (p. 119).

9   Neutra, op. cit., p. 20

10   For a comprehensive review of this issue, see Alberto Perez Gomez, *Architecture and the crisis of modern science* (MIT Press, 1983) and Alberto Perez Gomez, Introduction, *Ordonnance for the five kinds of columns after the method of the ancients / Claude Perrault* (Santa Monica, CA : Getty Center for the History of Art and the Humanities, 1993)

11   Neutra, Op. Cit. p. 32 *"Rationalism in science and social philosophy had evidently failed to find an adequate architectural expression of its own, and succeeded only in producing another fashion, one that was curiously divorced from the requirements of real life… The ideal of an integrated environment, which was to be the programmatic expression of a new order, seemed forgotten. …From Louis Philippe to Queen Victoria, the middle classes, political victors though they were, kept on rehashing and diluting the forms of the ancien regime that they had upset."*

12   Neutra, Ibid., p. 30. *"The common people had always been the majority; but only now did the majority become important: the new science, based on mathematics and measurements, seemed to implied the glorification of numbers. …Moreover, it was amply demonstrated by the new industrial technology that science really 'worked.' …. The mere fact that it did work and produce… was startling to the common man."*

13   Ibid., pp. 28–39.

14   Ibid., p. 53.

15   The influence of Immanuel Kant's *Critique of Judgment* (1790) on architectural theory is discussed by Adrian Forty in his essay on 'Form' in *Words and Buildings, A Vocabulary of Modern Architecture* (Thames & Hudson, 2000).

16   Ibid. Neutra, p. 54.

17   Ibid., p. 52.

18   Ibid.

19   Ibid.

20   Buckminster Fuller, "The Cardboard House", *Perspecta*, Vol.2 (1953), pp. 28–35.

21   Buckminster Fuller, "The Industrialized House, Incorporating Fuller's Geodesic Dome & Standard of Living Package," *Perspecta*, Vol.1 (1952), pp. 28–37.

22   C. Theodore Larson, "Toward a Science of Housing", *The Scientific Monthly*, Vol 6 (Oct 1947), pp. 295–305. See also Franca Trubiano, "Designing for low energy- seeking representations of high-performance homes in post war America" in *Architecture and Energy*, (Routledge Press, 2013).

23   Neutra, *Survival Through Design*, p. 51.

24   Neutra, p. 52.

25   Ibid., p. 56.

26   Ibid., p. 54, 55.

27   Ibid., p. 55.

28   Kevin Kelly, *What Technology Wants* (London Penguin Books, 2010). pp. 133–143. See Chapter 7, On Convergence, wherein Kelly describes the truly collaborative origins of the electric light bulb.

29   Neutra, op. cit., p. 55.

30   See Adrian Forty, *Words and Buildings – A Vocabulary of Modern Architecture* (Thames & Hudson, 2000), Part Two, on 'Function', pp. 174–195.

31   Ibid., p. 53.

32   Ibid., p. 19.

33   The US Census bureau see http://www.census.gov/dmd/www/resapport/states/california.pdf (accessed August 07, 2012).

34   *Nature Near, Late Essays of Richard Neutra*, (Capra Press, 1989), p. 76.

35   Neutra, op. cit., p. 16.

36   Ibid.

37   Ibid., p. 20.

38   Ibid.

39   Ibid., p. 21.

40   Ibid., p. 20.

41   Ibid.

42   Ibid., p. 82.

43   Ibid., p. 83.

44   Ibid., p. 84.

45   Ibid. p. 83.

46   Ibid., p. 20.

47   Ibid., p. 85.

48   Ibid., p. 86.

49   Ibid., p. 91.

50   Ibid., p. 92.

51   Ibid.

52   Ibid., p. 93.

53   Ibid., p. 4.

54   Ibid., p. 159. Further evidence of his macro/micro scales of design is noted in Neutra's description of Einstein's contribution to the modern space concept. Echoing, in part, Sigfried Giedion, Neutra states, "*To the lay person there may well be some strangeness in contemporary mathematical physics, in 'curved, expanding, and contracting space,' in the 'indeterminacy' or 'uncertainty' of motion, position, velocity. This strangeness comes largely from one circumstance: all those concepts have their concrete stage of action quantitatively outside of the normal human sensory range. Planning and designing within submolecular spaces is called chemistry and, on a still more elementary plane, nuclear physics. Here new ideas must fit new minimum magnitudes. On the other hand also the investigation of vast magnitudes and velocities on stellar levels, and of a correspondingly sized physical interaction, is, naturally, unfamiliar to the man in the street. All this commonly remains outside the realm of planners and architects as well as of those who use cities and houses on an ordinary macroscopic scale. Investigations far beyond the daily routine could not well demonstrate to a general public how essentially near the new scientific space concept has come to our very physiological foundations.*"

5

Lightweight, Impermanent, Recycled

Gernot Riether

INTRODUCTION

Small-scale temporary architectural injections into urban settings are a potentially appropriate approach to activating public spaces especially in a recession, when resources to realize expensive urban projects do not exist. The ephemeral nature of these projects also has the capacity to reflect fast changes in technology within the framework of a digital culture. Some cities are using these strategically located, temporary, architectural injections as a tool to create new kinds of public spaces or transform existing ones. For architects and urban planners, these small projects are certainly a welcome opportunity to challenge conventional approaches to constructing buildings or operating in a city.

Each year, the Serpentine Gallery in London commissions an internationally acclaimed architect to design a temporary pavilion for its lawn. Of a similar scope are the Burnham Pavilions in Chicago, designed to underscore the city's historic reputation as a center of architectural innovation. As distinct from the temporary pavilions that we see in world fairs, these recent pavilions enter into deliberate relationships with the city fabric by creating or transforming existing public places.

The funders for these projects vary from non-profit organization to museums to private or commercial owners. In some cases, projects are simply self-initiated by artists or architects. Projects might be defined as event spaces, such as the Museum of Modern Art (MoMA) PS 1 Warm Up events in Queens, during which the PS1 courtyard is redesigned in an annual competition for young architects. Other projects are temporary architectural interventions that primarily have a commercial goal, but still impact public space, such as the Prada Transformer, a pavilion in Seoul designed by Rem Koolhaas for fashion shows, exhibitions, movie screenings and concerts. Such projects infuse urban spaces with a technological landscape of constant change.

The projects are small enough to allow for more radical architectural and urban experiments that speculate about our expectations of public environments. The Prada Transformer can be picked up by a mobile crane and rotated to accommodate different functions. The courtyard at PS1 has thousands of visitors awaiting its redesign during summer weekends every year. The ephemeral nature and spontaneity of these projects suggest a more dynamic public space that is constantly changing and being updated. These short-term projects have the capacity to suggest long-term strategies.

While small-scale temporary projects may or may not be interpreted as a response by public places to digital culture, they most certainly allow architects to experiment with the latest tools at hand. For a number of pavilions, the main objective is to test the limits of opportunities for digital design and fabrication, testing new material systems in full scale and rethinking the conventions of construction by demonstrating the latest developments in material-orientated computational design, simulation, and production processes in architecture. Such objectives are evident in a series of pavilions from the Architecture Association (AA) in London and a series of pavilions at the Institute for Computational Design (ICD) at the University of Stuttgart. The aim for Achim Menges and Jan Knippers and their students at the University of Stuttgart was to develop a dynamic structure using the elastic qualities of plywood strips. The pavilion was exhibited at the University's campus.

Not long ago, experiments in digital fabrication were limited to the scale of small installations. During recent years, these projects have reached a scale that impacts public space. Pavilions, such as the ICD Pavilion, are large enough to define a public space, yet small enough to be disassembled or recycled effectively. Mobility entails certain conditions. Building without a permanent foundation means that projects need to be lighter, a need for efficient transport of the project affects the size of modules, and the requirement for deployability informs the details of joints. That said, sometimes a pavilion designed as a deployable structure becomes permanent because of characteristics arising from its mobility. Chanel's Mobile Art Pavilion, commissioned by Karl Lagerfeld and designed by Zaha Hadid, served as a travelling fashion and art exhibition that was displayed at public spaces worldwide. After being staged internationally, the pavilion was finally bought by the Institut du Monde Arabe. Designed as a lightweight temporary structure, the pavilion sat easily on the institute's entry plaza, a site otherwise impossible to build on. A more typical building would have required modifying the parking deck structure below the plaza.

THE AIA PAVILION AS A CASE STUDY

The AIA pavilion in New Orleans was an attempt to synthesize two disparate issues on a tight budget. The first goal was to activate pedestrian traffic in areas otherwise inaccessible or forgotten. The second goal was to develop and test an environmentally friendly material system for a new kind of lightweight structure. So far, the AIA pavilion might reasonably be said to share similar goals with any of the Serpentine pavilions or a Burnham pavilion. However, as a selected entry for an art installation, the budget added an additional challenge. The budget for the AIA Pavilion was $2,500, which is very low compared to the Burnham Pavilion, with a budget of $500,000, or the PS 1installation, with a budget of $50,000.

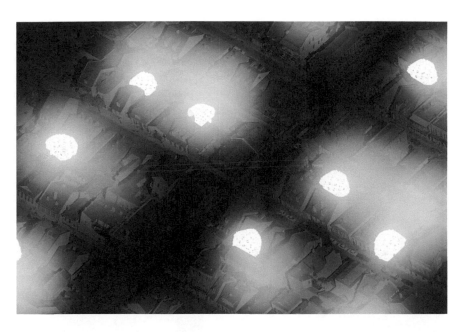

5.1   Placement of pavilions

The project was a response to an open call by the AIA New Orleans asking for interventions to bring to life the historic city of New Orleans. The proposed project suggested a series of pavilions that strategically occupy private courtyards and temporarily change them into public spaces. At night, the pavilions would dramatically modulate the host environment, bringing attention to the city's most romantic and mysterious spaces. These courtyards are usually hidden deep inside city blocks and are only accessible from the street through very narrow walkways.

In December of 2010, one of the pavilions was realized for the AIA's annual DesCours event. The pavilion was located at Orleans Street, close to North Rampart on the north edge of the French Quarter. From the street, a passerby could only see glimpses of the alien-like, bright, glowing object. For curious visitors, emboldened to enter the courtyard through a very narrow alleyway, the strange object revealed itself as an event space of open possibilities. The project was also published in local newspapers and a schedule of events at the pavilion was published at the DesCours website, which prompted people to search for the pavilion. Events that were scheduled in the evenings drew a larger audience to the pavilion.

The project's realization required the involvement of the courtyard's owner, who "lent" his space to the public for one week. The pavilion was developed as a modular system that could be assembled and disassembled during two days to allow for a fast construction process and an almost instantaneous change from a private condition to a public space. Developing the pavilion as a modular system was also necessary to respond to the difficult accessibility conditions of the site. The very narrow and angular alley from the street to the courtyard required small-size building components not exceeding a four-foot width or a seven-foot length.

In addition to these constraints, each module was designed to be light enough and small enough to be carried by one person. The modules ranging from one square foot to ten square feet in size could be easily stacked, which allowed one person to carry up to ten modules at a time through the narrow alleyway. Once all modules were in the courtyard, the entire pavilion could be assembled and disassembled by a single person. The modules were connected by standard bolt connections. Engraved edge numbers allowed for a self-guided assembly process without drawings or diagrams.

5.2   The pavilion at night

THE PAVILION AS A FLEXIBLE SYSTEM:

The pavilion was developed without knowledge of which courtyard would be available for the project. A final design was also needed to convince the owner of the site to participate. At the same time an important intention of the project was to make it site specific. This conflict was resolved by making the pavilion's modular system flexible enough to allow for adaptation to a certain range of site conditions.

A structural envelope was developed for a pavilion of 150 to 300 square feet. This envelope was made up of triangular modules. Each module was unique and was developed as a variation of a single cell. Each cell was generated from different sets of attributes that were derived from the cell's position within the overall form, different architectural and structural requirements and unique site conditions. Scripting the entire pavilion in Grasshopper, a graphical algorithmic editor, allowed for parametric transformation of each cell's geometry into functions of seating, foundation, light fixture, plant holder or rainwater collector. In that way, it was not only the overall form that could be adapted to a specific courtyard; it was also the module and type of module that could change based on a specific location. The number of modules that formed water collectors or were used as planting

5.3   Interior view of the pavilion

containers could vary based on site conditions. The planting container module itself could further vary to respond to different species of plants that might grow at a specific site. The final overall form and spatial qualities of the pavilion emerged from networking the cells into a multifunctional building envelope.

The pavilion was designed more as a flexible system than as a building with a fixed site, and the pavilion could therefore adapt to different sites. Generative modeling tools, in this case Grasshopper, allow for defining form-generative systems that can respond to unique requirements and contexts. Using such tools, form results or "emerges" from the definition of rules. This method of working suggests bypassing constraints of rigid architectural typologies.

During the design process for the AIA Pavilion, different geometric methods—such as nurbs, polygon and subdivision—were tested to develop new "digital typologies". The resulting building blocks or "cells" were tested against architectural functions, structural performance and materiality.

This liquidation of conventional types leads to more flexible systems, new combinations of functions, unexpected hybrid situations, new forms of interactions between natural and artificial systems and systems that can adapt to different needs. A certain level of complexity within the initial cell proved to be essential in allowing for the integration of programmatic, functional and contextual needs. Each cell in such a process developed from a primitive to a complex state.

The final base geometry of the pavilion's cell was a triangle, which transformed into 320 unique variations based on different sets of attributes. Depending on its position, the edges of each cell were folded differently to provide stiffness within the cell and to contribute to the overall structure. A single cell, for instance, could sometimes act more skin-like and sometimes more column-like by changing in shape and configuration.

The complexity of the cell's digitally driven typology allowed for combining the structure and envelope in a hybrid system. The edges of each cell were folded differently based on each cell's location within the overall structure. This provided stiffness within the cell. A complex geodesic system was formed that connected the edges of all cells. To minimize the amount of material used for the envelope and to create a lightweight structure, the envelope generated wormholes that acted like braces and columns. This formation of wormholes within the surface increased the surface tension, which stabilized the structure of the pavilion. This allowed us to keep the weight of the structure at 260 pounds and, simultaneously, keep the expense of the pavilion at $2,500.

The process of generating the form of each cell was based on information derived from the context and specific functions. It was a search for mathematical laws that natural and artificial systems might obey: a working concept of emergence that uses the mathematics and processes that are useful to designers. The process was entirely scripted, that is, articulated as rules that informed the geometric transformations of the cells. The degree of unpredictability of the final geometry of the cell increased with the increasing complexity of information that operated on the cell geometry. The final overall form and spatial qualities of the pavilion emerged from the cell's variations. Scripting the entire pavilion in Grasshopper

Flexible Mold

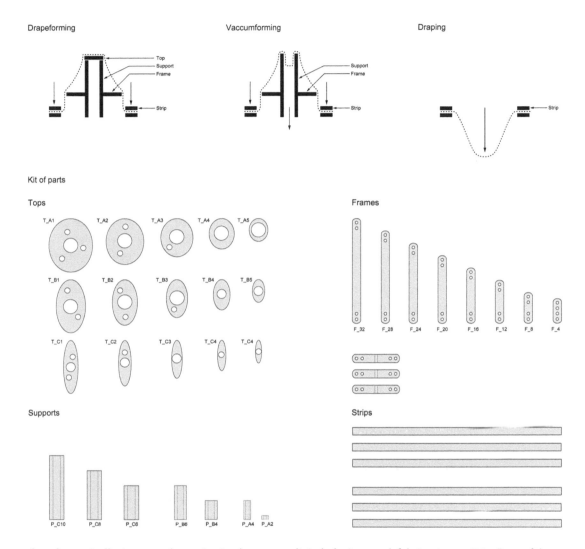

also dramatically increased continuity between digital design and fabrication. Numbering all edges and adding more than 2,000 connection details that varied based on cells' differentiation could be easily added to the script as an additional parameter.

5.4   Parts of the flexible mold

FLEXIBLE MATERIAL PROPERTIES

Developing a building system without certainty of its final form and function also informed the way of thinking about materials. Architects are accustomed to working within the fixed properties of traditional building materials, like stone, clay, wood, concrete, steel and glass. Collaborating with chemists might allow architects to work at the scale of the molecule and to reorganize the underlying structure of matter in architecture. Polymers, for instance, can be developed with specific properties from the bottom up rather than being shaped from the top down. This would suggest a more parametric material. Parametrically engineering materials such as polymers would yield a wide range of properties from a single material. When chemistry first came up with the innovative use of polymers in the 1960s, plastic was an environmentally problematic material. Plastic is celebrating a come back as the chemical industry moves away from fuel-based polymers and toward biomolecule-based polymers.

The AIA Pavilion was a part of a larger ongoing investigation into new techniques and methods to reintroduce plastics as legitimate building materials. The highly malleable nature of plastic was used to respond to the digitally derived form of the pavilion and its complex geometry and cell variations. Plastics are light, impact-resistant and easy to fabricate. Glycol-modified polyethylene terephthalate (PETG), a material that is fully recyclable, was used as the sole material in the system. The pavilion can be also seen as an attempt to convey new spatial and aesthetic qualities that may be essential in changing our perception of plastic from that of an environmentally-problematic to that of an environmentally-friendly material.

FLEXIBLE FABRICATION

Architecture systems need to be flexible enough to respond to parameters that are unique to each project and each site. Flexibility in manufacturing usually means the ability for one machine to produce different products or parts, the viability of an assembly process and sequence, or the ability to adapt to changes in the design. In industrial design, the term "Machine Flexibility" is used to describe machines that can manufacture a variety of products. The term "Flexible Manufacturing Systems" or "FMS" is used when several machine tools are linked in a flexible dynamic way.

Computer-Aided Manufacturing (CAM) processes like laser cutting, Computer Numerical Control (CNC) milling, and 3D printing, coupled with Computer Aided Design (CAD) tools and parametric modeling software provide instant feedback in a design process, which allows the designer to speculate with real materials in real time. This integration of production processes as active agents in the development of architectural systems suggests a design process that is driven by the incorporation of all possible parameters and the analysis of the consequences of their interactions. Mainstreaming multiple processes usually takes a lot of effort at the front end that will pay off in mass customization, a development of unique variations based on functional and contextual changes for different users and different sites.

In the pavilion, design and fabrication use the same software platforms that helped streamline the production process. The connection between digital design and fabrication has become obvious and natural. Because the parametric model was entirely scripted, the project was linked to fabrication parameters and a flexible manufacturing system. A flexible mold was developed that could produce different shapes from different triangles. The mold was constructed from a digitally fabricated kit-of-parts that combined three different thermoforming techniques: drape forming, vacuum forming and draping. This flexible system responded well to a parametric digital model to fabricate the cells. Because the pavilion was entirely scripted, the model could instantaneously react to fabrication constraints at any time. A continuous feedback loop was created between digital modeling and fabrication.

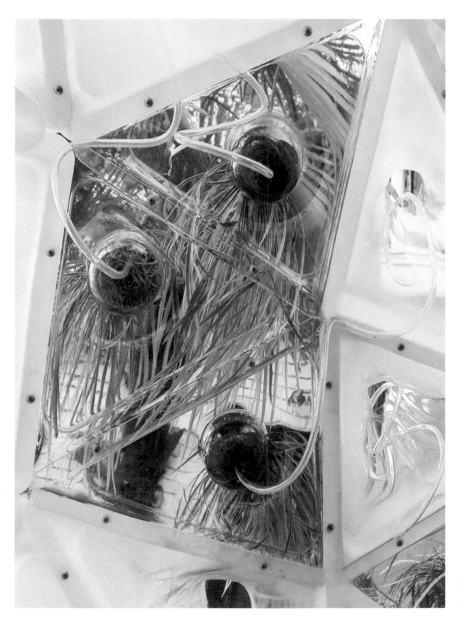

5.5   Plants inhabit the pavilion's envelope

5.6　Plants inhabit the pavilion's envelope

Developing our own flexible mold used fewer parts and tooling than other molding techniques, which allowed us to save material and for cost- and time-efficient production. It was also very precise, resulting in very small assembly tolerances. After production, all prefabricated cells were assembled into larger components that were designed to stack and fit compactly into a small truck for transport.

It was not only the overall form that can adapt to a specific courtyard, it was also the module and type of module that can change based on a specific location. The number of modules that were water collectors or the number of modules that were planting containers for instance varied by location. The parameters of the site were used to inform the overall form, but they also differentiated modules within the pavilion's envelope.

CONCLUSION

The AIA pavilion project was only possible by developing a new relation between an architecture organization, a city, a private residence and the industry. The project was published online and has been discovered by collaborative and non-profit organizations in other cities who are interested in the idea of a pavilion as a temporary architectural structure that revives street life. More recently, a variation of the pavilion has been installed in Chattanooga, Tennessee. Here, the pavilion will travel between different empty lots in downtown Chattanooga to draw attention to otherwise abandoned urban places. Such an approach implies only temporary change, but has the capacity to suggest otherwise impossible changes and immaterial connections within the urban fabric. It is also an affordable response for cities that are suffering from the current financial crises and often have limited budgets at hand. The light and temporary quality of the pavilion allowed it to occupy places that would otherwise be taboo for public use, an architectural Occupy Movement to test out urban ideas and to make cities more lively. In doing so, the AIA pavilion promotes an architecture that avoids typological thinking, suggesting single systems that can be complex and flexible enough to respond to multiple functions and contexts. The project privileges the design of flexible systems rather then fixed buildings. In that sense, it suggests mass customized building systems and system components that allow for more intense relationships between architecture and its various environments.

The project suggests an aesthetic quality that emerges as a consequence of interactions, matter and material behavior. Coupling material research with scripting as a design technique and a flexible manufacturing system allowed customizing each cell according to different functions and contexts in a highly cost-effective way. The pavilion presents a new form of lightweight structure that saves on building materials. The project also recognized recent trends in the chemical industry that suggest an architecture built from recyclable bio-materials. The project further suggests the possibility of manipulation of material at a micro scale. Applying this idea to architecture might suggest a very different future for architectural matter, an architecture that is responsive to its environment in a much more dynamic way. Buildings might be made from cells that can be manipulated to alter their architectural meaning in the search for equilibrium between architectural function and a constantly changing environment.

ACKNOWLEDGEMENTS

The AIA pavilion was realized as part of a graduate options studio that Gernot Riether taught at the Georgia Institute of Technology. The team of students supporting this project was Valerie Bolen, Rachel Dickey, Emily Finau, Tasnouva Habib, Knox Jolly, Pei-Lin Liao, Keith Smith, and April Tann. Consultants Russell Gentry provided structural specialization and Andres Cavieres provided technical support. The fabrication took place at the Digital Fabrication Laboratory (DFL) at the Georgia Institute of Technology. Special Thanks to the American Institute of Architects (AIA) in New Orleans, and the School of Architecture at the Georgia Institute of Technology.

REFERENCES

Braskem. http://www.braskem.com.br

Dow Chemical Co, "Dow Brazil: Scaling New Heights," http://www.dow.com/ PublishedLiterature/dh_0191/0901b80380191026.pdf?filepath=news/pdfs/ noreg/162-02356.pdf&fromPage=GetDocIbid., accessed July 12, 2011.

Lorcks J., "Bio plastics: Plants, Raw materials, Products". Fachagentur Nachwachsende Rohstoffe. German Federal Ministry of Food, Agency for Renewable Resources, accessed July 12, 2011, http://www.fnr-server.de/cms35/index.php?id=200

M.M. Hanczyc, T. Toyota, T. Ikegami, N. Packard and T. Sugawara, 'Fatty Acid Chemistry at the Oil-Water Interface: Self-Propelled Oil Droplets', *Journal of The American Chemical Society* 129(30): 2007, pp. 9386–91.

Menges, A. 2011a. *Integrative Design Computation: Integrating material behavior and robotic manufacturing processes in computational design for performative wood constructions.* ACADIA 2011: Integration through Computation. Banff, Alberta, 72–81.

Menges, A. 2011b. *Integrative Design Computation: Integrating material behavior and robotic manufacturing processes in computational design for performative wood constructions.* ACADIA 2011: Integration through Computation. Banff, Alberta, 77–81.

Riether, G. 2011. Adaptation: A pavilion for the AIA in New Orleans. ACADIA 2011: Integration through Computation, (Projects). Banff, Alberta, 52–57.

Riether, G. 2012. Pavilion for New Orleans. DETAIL /2012 ('Vorfertigung') pp. 61, 613, 642–644.

Riether, G., Knox, J. 2011. Flexible Systems: Flexible Design, Material and Fabrication: The AIA pavilion as a case study. eCAADe 2011: Respecting Fragile Places. Ljubljana, Slovenia. 628–634.

The Renewable Corporation, http://www.therenewablecorp.com/eco_efficiency/ sugarcane_as_feedstock.htm, accessed July 12, 2011.

U.S. department of energy. Biomass Program, direct Hydrothermal Liquefaction. Energy Efficiency and Renewable Energy. http://www1.eere.energy.gov/biomass/

Weinstock M., "Morphogenesis and the Mathematics of Emergence", in *Theories and Manifestos, of Contemporary Architecture*, Jencks C., Kropf K. (eds), 2006, Chichester, West Sussex, Wiley.com, p. 273.

6

On Enterprising Architecture

Nathan Richardson

> *Even a brief conversation with architects these days soon reveals their concern about the future of the profession. Two conditions seem to have generated the air of crisis. The first is the rise in the number of professionals who now are unemployed, compared with three or four years ago ... Architects are going about wondering whether the market for their services will ever [return].*[1]

Although the preceding commentary will no doubt resonate with those in the profession that have experienced the decline brought on by the 2008 financial crisis, it wasn't written in that context. It is from an essay penned over thirty years ago by Robert Gutman in response to the severe economic recession of the mid-1970s. It was a call to embrace an entrepreneurial architecture, "one more aggressive in getting work and creating its own demand."[2] Gutman argued that entrepreneurship was essential in order to respond to a number of significant issues. First, much of an architect's work could be outsourced to other professions. Additionally, the demand for architectural services—with its dependence on construction and development—was uniquely sensitive to fluctuations in the economy. Furthermore, the architect's influence among the sphere of stakeholders within a given project was eroding. The fact that the profession is now suffering the same circumstances for many of the same reasons (and has a number of times since the 1970s) initially appears to support one of at least three conclusions. First, one might suggest that these recurring conditions are beyond the control of the profession and though difficult, remain sufficiently endurable within standard forms of practice. The inadequacy of such a call to inaction appears self-evident. Second, one might argue that the profession has embraced entrepreneurship, and found itself no less exposed to the difficulties brought on by the more recent economic recessions. This conclusion doesn't seem supported by any broad and substantive shift in architectural practice over the last half century, though the profession has naturally evolved. Such an evolution in practice and pedagogy is discussed in Joan Ockman's recent volume on architectural education. In it, George Johnston notes the increased heterogeneity of project participants, delivery methods, and professional identities over the last 40 years[3] and Mark Linder notes a similar trend in the 1990s: "Instead of adapting concepts and discourses from other fields, architects undertook new kinds of opportunistic projects, embracing instrumentalities and logics that both intensified their professional expertise and expanded the discipline's reach."[4] Significant as these changes are, it's not clear that architects have embraced entrepreneurship in a way that substantially affects their business models. This leads to a third and more plausible conclusion. That is, entrepreneurship can have a significant impact within the profession, and though a few have embraced it, the profession writ large doesn't have a substantive understanding of what constitutes entrepreneurship nor a rigorous framework for exploring its potential impact within their practice.

The following research is more than a call to entrepreneurship; it is an exploration of the most proximate form of entrepreneurship to conventional practice: entrepreneurship as architecture and architecture as real estate. It is a study of past and current practitioners of architecture that have engaged in a broad range of real estate activities as entrepreneurs. An understanding of this form of entrepreneurship and its relevance in today's context necessitates a review of precedent, an exploration of practice, and a consideration of the prospect for such activities. If architects develop a more substantive knowledge of this form of entrepreneurship in architecture, they may find enhanced viability in the face of persistent uncertainty. First however, an exploration of what constitutes architecture and entrepreneurship is in order.

ARCHITECTURE AND ENTREPRENEURSHIP

Architects often frame their professional identity with almost exclusive respect to the buildings they design. This identity manifests itself in the profession's formation of history, education, training, and practice. In reality, few architects have ventured far from a common conception of practice in which they provide design services to a client who intends to build. If one wants to find a definition of architecture that suits their objectives, there is plenty of material from which to draw a well-nuanced version that fits. However, architecture in many cases is framed as a critical societal, cultural, artistic and/or professional production in which the architect plays a central role. As Andrew Saint argues in *The Image of the Architect*,

> *Down the centuries one strain of architectural ideology has been heard much louder than others. That is the strain of artistic individualism, which ascribes both merit in particular buildings and general progress in architecture according to a personal conception, usually of style, embodied in buildings and developed from architect to architect over the course of history.*[5]

This view only adds a degree of autonomy to the idea that architects, at a fundamental level, do little more than design buildings. As Spiro Kostof explains, "… this is what architects are, conceivers of buildings … The primary task of the architect, [in antiquity] as now, is to communicate what proposed buildings should be and look like."[6] Historically, such a model of practice—while dominant—is far from singular. In the future, an alternate view of practice may find heightened traction in the face of emerging challenges and increased uncertainty.

Consider the following: "Entrepreneurship is a process by which individuals … pursue opportunities without regard to the resources they currently control."[7] Although this definition emerged from a business-oriented body of research, it bears a striking resemblance to the activities of an architect. In other words, architects are adept at pursuing opportunities to shape the built environment without much deference to their relatively limited control of the capital resources employed in building. Another commonly cited definition of entrepreneurship frames it as the process of creating value by bringing together a unique combination of resources to exploit an opportunity.[8] This statement can likewise be understood in the context of architectural practice; architects are no doubt skilled at leveraging opportunities by bringing together a diverse combination of resources in construction to create value in architecture. Simply put, every architect is an entrepreneur in some venue. Even though architecture can be understood as an entrepreneurial endeavor, entrepreneurship isn't often an explicit part of architectural practice or education. As such, architects rarely describe themselves as active entrepreneurs or leverage their entrepreneurial potential in venues other than conventional practice. Clearly, the built environment and those that design it are critically important, but there is a broader array of methods by which architects can make meaningful contributions to society. Even though the potential application of such an entrepreneurial architecture is broad enough to transcend the discipline entirely, most precedent for entrepreneurship within the discipline grew incrementally as an extension into related segments of the real estate industry.

The entrepreneurs that are central to the following research are significant in that they leverage an uncommon expertise in real estate and embrace a broader range of roles than those conventionally associated with an architect. First and foremost, these entrepreneurs are architects. While the examples cited below span three centuries of time, each is considered an architect—or designer of buildings—

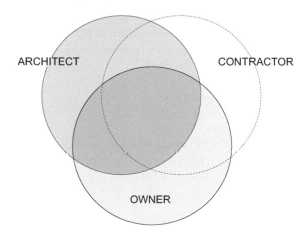

6.1   The integration of otherwise distinct roles, such as *owner* and *architect* are central components of this research, though the role of *contractor* is often integrated

according to the standards of their day. What makes them entrepreneurs, however, is the degree to which they've embraced the larger project delivery enterprise through ownership or some form of direct investment in speculative projects. In addition, these architects were substantially involved in the conception of these projects at their earliest stages. Some may also integrate various forms of engineering and many may integrate construction, but such integration alone doesn't constitute an exceptional illustration of entrepreneurship in the context of the following research. When architects combine real estate expertise with design practice, they can approach architecture in a particularly holistic manner, incorporating all its procedural and characteristic complexities: economics, finance, project delivery, marketing, and operations. These divergent models of practice demonstrate alternative avenues that, while imposing additional operational risk, present the potential for enhanced viability in an age of economic uncertainty. Given the real estate market's role in the recent economic crisis, this intellectually expansive model of practice is ripe for critical exploration. The current uncertainty faced by the profession of architecture, the development industry, and society in general warrant a more broadly informed architectural profession than that currently constituted.

Learning about this form of entrepreneurship requires setting aside the common understanding of both architect and real estate developer. It is often held, in light of common stereotypes, that the designer is an image-driven artist primarily concerned with the appearance of a building and minimally concerned with the financial performance of the investment. It is also held that the real estate developer is a financially-driven profiteer with little care for design and architectural innovation. In this oversimplified dyad, the designer wants to do what has *never* been done *no matter* the cost and the developer wants only to do what *has* been done at *minimal* cost. If these hold true, there is little common ground between these centers of knowledge and little opportunity for a productive integration of roles. However, these are counterproductive stereotypes that many in the development, design, and construction industries can move beyond. This is evidenced through widespread collaboration on a *project-by-project* basis and more critically, through successful forms of integration at a *practice* level. Nevertheless, many in the profession of architecture are relatively uninformed about the degree

6.2 Selection of historic architects engaged in real estate oriented entrepreneurship, representing a thread of architectural history often left untold

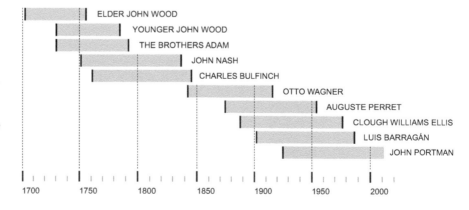

to which architects have integrated real estate and entrepreneurship in their practices. Most can cite the work of John Portman, but a deeper understanding of other relevant precedents rarely follows. In matters of design, architects frequently leverage their knowledge of history in shaping their future. A similar exploration is no less important in the context of entrepreneurship.

PRECEDENT

The integration of architectural practice and real estate enterprise extends well into history, though it isn't an established part of the architecture profession's historical narrative. Among the ranks of enterprising architects is John Wood the Elder and Younger, The Brothers Adam, John Nash, Charles Bulfinch, Otto Wagner, Auguste Perret, Luis Barragán, and John Portman.[9] The careers of these well-known figures offer some insights into the operations of entrepreneurial architects. First, one can see the broad range of capacities in which these architects engaged real estate development. John Wood the Elder for instance, initially functioned as an agent for his client (landowner Robert Gay) and then moved into a primary role by securing a land lease to develop Gay's property.[10] Robert and James Adam were directly involved in the conception, financing, construction, and promotion of their street schemes. Their Adelphi Terrace is an excellent example of the consequences of assuming additional risk in the storm of a financial crisis not too dissimilar from America's in 2008.[11] John Nash functioned as an agent, investor, and advisor to clients and the crown during his professional career. While Nash also engaged in ownership, most of his experience leads one to consider a more expansive understanding of the integration of real estate, primarily advisory in nature, based on specialized expertise. Charles Bulfinch and Luis Barragán both formed partnerships to execute their developments, with varying degrees of success. Sir Clough Williams Ellis and John Portman operated to conceive, design, develop, and even operate the business that would lease their buildings. Otto Wagner occasionally conceived projects and promoted them for potential clients to embrace, but more often conceived, constructed, financed, and leased his own projects. Auguste Perret and his brothers were builders, designers, and investors in apartment buildings that would, in some cases, also serve as their residence and offices.[12] These precedents represent nearly three centuries of this form of entrepreneurship in architecture. While the profession might find more recent models particularly relevant, the long history of entrepreneurship within the profession (and its absence from the typical historical narrative) speaks volumes.

The range of activities represented by these historic precedents, particularly in the realm of real estate, hasn't changed considerably as one considers the work of contemporary architects. Although this form of entrepreneurship is sometimes difficult to identify and document in contemporary practice, this research studied the real estate activities of about thirty architects. Among the pool of current precedents are Bruner/Cott, Randy Brown, David Nielson, Campbell Ellsworth, Perkins Eastman, KRDB, Tom Allisma, Sebastian Mariscal, and Jonathan Segal.[13]

These practitioners have engaged in the conception, design, and delivery of speculative real estate projects (often as owner/investors) at some point in their careers. Simeon Bruner and Leland Cott of Boston, Massachusetts leveraged real estate development early in their firm's history as a strategy to establish their architectural business. While the firm no longer develops properties, ownership and property management remain a key part of their organizational success. Similarly, Campbell Ellsworth of Cambridge, Massachusetts, KRDB of Austin, Texas, and Jonathan Segal of San Diego, California have engaged in some form of direct investment in real estate. For these practitioners, this model has allowed them more discretion in taking on certain types of work. Other practitioners, such as Brad Perkins of Perkins Eastman and David Nielson of Boston, Massachusetts have utilized real estate expertise to offer a distinct set of services to their clients and, in some cases, share additional risk.

As one synthesizes these precedents from a series of isolated historical artifacts in search for broader themes, a series of concepts emerge. First, a look at these practitioners in the aggregate suggests a sequence of prudent steps: integrate expertise, form strategic partnerships, and take on investment risk when necessary. This sequence is particularly important as one considers an extension into real estate as a part of the current economy. Opportunities are certain to exist, but they need not all require direct investment; expertise should certainly come first and partnerships may be preferable to direct investment. Second, their method of operation is largely necessitated by their values. Beyond financial incentives, each of them aims to bring something distinctly valuable to the marketplace. One can make a compelling argument that these practitioners often meet needs or spur demand that often go unaddressed through standard channels of project delivery. By doing so, many of these practitioners were better able to create exceptional value within the marketplace. Third, one must acknowledge that the integration of such diverse roles can lead to a complicated set of identity conflicts and ethical challenges. For instance, John Nash declared bankruptcy as a tradesman because such a course of action was unavailable to those with professional standing, such as architects.[14] Charles Bulfinch experienced similar difficulty in negotiating the realities of bankruptcy with his public role as an architect.[15] More recently, John Portman was questioned by the AIA regarding the potential conflicts of interest related to his integrated activities.[16] Fourth, these professionals and their work demonstrate what should otherwise be known to the industry: "design innovation" and "investment performance" aren't inherently at odds. Some of the most transformative works of the design profession, which receive due attention from architects, were speculative buildings conceived, developed, and built by entrepreneurial architects.[17] While this research doesn't pretend to sufficiently analyze the architectural quality of their work or quantify the precise nature of their profit, it does acknowledge the architectural contributions made by these architects while simultaneously dealing with financial constraints.

Without engaging in an extensive analysis of the design merit of their work, one can acknowledge that many of these architects have made significant and lasting contributions to the advancement of the discipline's oeuvre. This is evident enough if one considers the persistent place architects like John Wood, John Nash, Charles Bulfinch, Otto Wagner, August Perret, and Luis Barragán (a Pritzker Architecture Prize winner) have in the historical canon of architecture. Though many of the contemporary examples can't yet make such a claim, they can cite a noteworthy set of regional and national design awards granted by their professional organizations; Randy Brown and Jonathan Segal are among the most prolific. Furthermore, most current practitioners do note their ability to deliver a product more consistent with their vision than they could through traditional channels of practice. As such, these architects provide an indication of the potential for substantive success through some form of integration. As these issues are sifted from a broad review of historic and contemporary precedents, it is necessary to consider a framework which can better inform the prospective practice of enterprising architecture.

Even a cursory review of these precedents yields considerable insight into the future viability of such forms of entrepreneurship. The following conclusions are particularly noteworthy. First, the extended history of these activities alone points to their usefulness in evolving and uncertain environments. Second, the diversity within this subset of entrepreneurship point to a range of models which could comprise an effective strategy for a highly adaptable architectural organization. Third, the fact that so many practitioners have found success through an expansion of architectural practice points to the potential success for those that embrace it in the future. Finally, it is also evident that many of these entrepreneurial activities brought enhanced risk to their practitioners (risk that played out as near financial ruin in certain cases). These conclusions point to the need for a more substantive understanding of these models of practice and the activities they may accommodate.

PRACTICE

There are a number of ways to envision, frame, or understand the diverse activities that constitute entrepreneurship in architecture. One could consider the primary stakeholders in a given project: owner, architect, and contractor (as in Figure 6.1). This framework however is burdened with the structure of conventional roles in project delivery, failing to completely address the diverse activities associated with each role, the implications of integrating roles, or the practice-oriented issues an architect confronts as an entrepreneur. Another, perhaps more relevant framework

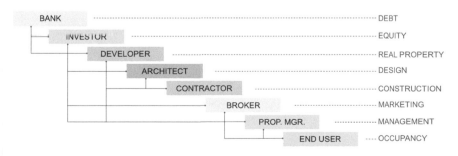

6.3 Design and development value chain with key parties, contractual relationships, and key products or services noted

for understanding entrepreneurship in real estate is represented through the value chain.[18] The emphasis here is on the production of architecture as an extended real estate process, rather than a few key roles in design and project delivery. It effectively communicates the broad range of activities and stakeholders that contribute to the formation of the built environment. However, this framework doesn't fully capture the key issues architects have confronted as they expand the entrepreneurial nature of their practice.

As a more effective alternative, the following framework of practice is proposed to organize specific models of entrepreneurial architecture with their most relevant and determinant characteristics. It stands as a more critical way to classify the broad range of activities embraced by the preceding practitioners. In short, these professionals are identified by the level to which they integrate *expertise*, *risk*, *resources*, and *identity*. These critical characteristics form the basis for the following distinct models of integrating architecture, entrepreneurship, and real estate.

The *service* model identifies a firm that still relies on fee-for-services income, but has internalized a level of expertise in real estate development which differentiates it from traditional architectural counterparts. They are entities primarily engaged in offering consulting services—architectural or otherwise. As such, they are more dependent on client-originated work. The *alliance* model describes a firm that retains real estate expertise, but also takes on certain project risks through partnerships. This can be accomplished by exchanging fee-based income for an equity stake in a given project or placing some portion of an architectural fee at risk, pending certain project outcomes. Such partnerships are typically structured

6.4 Models of integrated architecture and real estate enterprises, indicating their determinant characteristics

SERVICE	ALLIANCE	MULTILATERAL	UNILATERAL
			IDENTITY
		RESOURCES	
	RISK		
EXPERTISE			

6.5 Models of integrated architecture and real estate enterprises, indicating their relative characteristics and implications

	SERVICE	ALLIANCE	MULTILATERAL	UNILATERAL
INCOME BASE	FEE ONLY	FEE PRIMARY	FEE/EQUITY	EQUITY PRIMARY
RISK/RETURN	LOW	MODERATE	HIGH	HIGH
AUTONOMY	LOW	MODERATE	MODERATE	HIGH
PRODUCT DIVERSITY	HIGH	MODERATE	MODERATE	LOW
GEOGRAPHIC RANGE	HIGH	MODERATE	MODERATE	LOW
FIRM SCALABILITY	HIGH	HIGH	HIGH	LOW
FIRM ADAPTABILITY	MODERATE	MODERATE	HIGH	HIGH

on a project basis. The *multilateral* model describes a group of organizations that collectively share expertise, risk, and organizational resources while acting in concert as collaborative entities. These organizations often share financial, intellectual, or other operational resources. The *unilateral* model describes the organization that most fully integrates expertise, risk, resources, and identity, retaining the ability to act independently on projects as a single hybrid organization.

There are tradeoffs to be considered in pursuing certain models over others, and understanding them is particularly important in the face of today's uncertain economic conditions. Among the key characteristics of these distinct organizational models are *income base*, *risk/return profile*, *autonomy*, *product diversity*, *geographic range*, *scalability*, and *adaptability*. First, the multilateral and unilateral models engage a higher *risk/return profile* as compared to other models. Likewise, their compensation is based in equity income as opposed to fee-for-services income. The *autonomy* concept recognizes the ability of an organization to act independently as it has internalized a broad range of knowledge and activities and operates with relatively less dependence on other firms. *Product diversity* reflects the reality that a firm of a given size will find it more difficult to engage in a broad range of building types, while at the same time embracing a broader range of roles in the process. Conversely, firms with a limited range of roles within the industry (such as architectural or real estate services only) are better able to embrace a diversity of building types. The same concept holds true for *geographic range*, which reflects the estimation that a firm of a given size will have a greater ability to participate in projects across a broader geographic area when their activities are limited in services, such as a more pure form of architectural practice. Alternatively, the highly integrated models that take on a broad range of in-house activities such as development, design, construction, and operations, limit their geographic reach. There are certainly exceptions to these tradeoffs, but the practitioners studied in this research largely support the issues matrix. It is also recognized that firms of any substantial size more commonly exhibit the *service*, *alliance*, or *multilateral* models, whereas the *unilateral* model is commonly exhibited by small firms. Although this research doesn't extensively address the concept of *adaptability* over time, it is estimated that those firms that incorporate a full range of activities are unique in their ability to adapt to a given project or changing economic climate. All of these models are considered more adaptable than conventional architectural counterparts, given the additional expertise they can leverage in response to the changing needs of the economy, their clientele, or a specific project. This research uncovered anecdotal evidence among practitioners to support this claim, particularly in the wake of the recession following the 2008 financial crisis. It is important to remember that while these models are drawn from the preceding historic and contemporary models of practice—and do provide meaningful insight into the distinct ways that one may organize an integrated real estate and architecture business—the distinctions aren't always bright and pure. Firms and practitioners can find themselves moving across these boundaries over time or on a specific project for any number of reasons. Practitioners should understand the implications of each model in order to make informed decisions about the way they integrate, operate, and market their organizations.

Understanding the importance of precedent and the models of practice they support, doesn't entirely address the prospect of such models in today's context or that of the future. A substantive review of precedent should do more than point to what has been done and how it was accomplished. Such knowledge of precedent and practice is essential, but it stands as an incomplete representation of *why* an architect may choose to pursue a real estate expansion of their architectural practice and *how* it might be effective in accomplishing their larger and more substantive objectives.

PROSPECT

In any organization, it's critical to set a vision and define its purpose, mission, or fundamental objectives in order to appropriately implement the knowledge, resources, and activities that will form its constitution. Otherwise, one couldn't know that an appropriate plan was in place or be entirely sure how to measure its effectiveness. In order to ground such a discussion in the discipline of architecture, consider the following vision set out by the American Institute of Architects (AIA). The AIA was established in 1857 to "[promote] the cohesion of the architecture profession and [enable] architects to better serve their clients and improve the built environment."[19] What component of this purpose statement will best serve the profession in the increasingly uncertain times the profession faces: professional cohesion, client service, or improving the built environment? It's the latter which most succinctly reflects the principal objective for many of the practitioners studied here. Rather than embrace professional cohesion or client service exclusively, they leveraged an entrepreneurial process in order to improve the built environment. In reality however, their objectives are more nuanced. Entrepreneurship in architecture and real estate represents a particularly valuable opportunity to accomplish some of the following objectives. Clarifying these objectives will allow the practitioner to more suitably assess alternative models of practice that are most appropriate for accomplishing their aim. Among them are *environmental*, *economic*, *urban*, *social*, *theoretical*, and *typological* objectives. The shear diversity of goals exhibited by these architects is some indication of the potential effectiveness of this form of practice given the commensurate diversity of issues that marks our present and uncertain age.

Environmental Objectives

Naturally, issues of sustainability are at the forefront of professional practice. The construction and operation of buildings is commonly cited as a key contributor to energy consumption and greenhouse gas emissions. Current and past practitioners have found a greater level of effectiveness in addressing issues of sustainability through this integrated form of practice.[20] Firms that capitalize on the integration of design, development, construction, and property management have an opportunity to expand their sphere of influence in an effort to curb energy consumption and emissions across the life-cycle of the buildings for which they're increasingly responsible.

Economic Objectives

It's not an entirely new suggestion that architects engaging in real estate development may function as some part of the answer to making housing more affordable. In the 1970s there was a call for an integrated solution to the production of low income housing.[21] For reasons that can't be fully addressed here, it doesn't appear that a significant number of architects have pursued an integrated model with this objective in mind. However, the work of Bruner/Cott in New England should be cited as a particularly successful model of this opportunity, as they were able to implement these goals in their adaptive-reuse projects.[22] Given still more recent calls for solutions to affordable housing issues and the profession's role in addressing them, it appears the opportunity still exists, however challenging it may be.[23] KRDB is one example of a current practitioner that has grounded much of its work in an effort to make "design" more broadly affordable.[24]

Urban Objectives

Urban issues were particularly important to John Portman as he held convictions about the value of development in the "coordinate unit pattern."[25] Although he saw multiple avenues for addressing the poor patterns of development, he chose to take an entrepreneurial approach. He engaged in an integrated practice in an effort to find expanded opportunities to implement his architectural and urban vision. While he found a high degree of success in implementing his ideas, many have criticized his work for its entrenched disengagement from the broader urban fabric.[26] Whether or not Portman had the right vision or set of solutions, he certainly had a more significant impact on the urban condition through his form of entrepreneurship. One can point to any number of sub-optimal conditions in the current patterns of urban development and architects may likewise find greater opportunities for impact through an integrated model of practice. Even addressing these issues as an architectural practitioner requires a more rigorous understanding of the underlying issues that reside in the realm of real estate.

Social Objectives

This objective is centered more specifically on positively impacting people and communities directly. Given the professional standing of architects and their responsibility to the public at large, this is a particularly relevant objective for practitioners to actively embrace. Consider the following section of the AIA Code of Ethics: "Members should render public interest services, including pro bono services, and encourage their employees to render such services ... including those rendered for indigent persons, after disasters, or in other emergencies."[27] Of course, many architects and firms are already engaging opportunities within the traditional conception of practice to these ends. However, expanded opportunities are almost certain to be found through entrepreneurship in real estate. One particularly compelling precedent is the work of John Peterson through Public Architecture. He expanded his model of practice in an effort to initiate projects to benefit his community in San Francisco.[28] In many communities, cities, and countries there are real and immediate social needs that may be addressed more effectively through such an entrepreneurial approach to architectural practice.

Theoretical Objectives

Acting on a particular theory about architecture may be the most prevalent objective for many of the practitioners engaging in entrepreneurship. Historically, Otto Wager, Sir Clough Williams Ellis, John Portman, and Luis Barragán demonstrated their ability to implement deep theoretical convictions in their speculative work. Campbell Ellsworth, Jonathan Segal, and Randy Brown are a representative sample of current practitioners that have a similarly focused purpose in their activities. For many practitioners, they simply don't find commensurate opportunities to explore their theoretical values within typical channels of project delivery. This is especially true in the purely speculative residential market, where limited architectural innovation has long persisted.

Typological Objectives

Typologies in the residential sector are more commonly addressed than those in any other sector by these entrepreneurs. It is attractive for many reasons, not the least of which is its relative affordability and accessible scale. An architect's insight in typological innovations doesn't always coincide with their traditional clientele, making such an opportunity more fitting for speculative exploration. There are many typological limitations associated with speculative investment however. Among those generally outside the purview of a highly integrated practice are certain institutional projects such as churches, libraries, schools, and museums. This alone, may be sufficient cause for an architect to retain their traditional role as a consultant to a client. However, for those with particular insights into the residential, commercial, retail, and hospitality sectors, speculative entrepreneurship may be a worthwhile pursuit. Of note is Perret's pioneering use of reinforced concrete across building types, John Portman's hotel atrium innovations, Bruner/Cott's innovative work in residential adaptive re-use, Tom Allisma's retail specific design and development, and KRDB's effort in pre-fabricated residential architecture.[29]

Although these underlying objectives demonstrated by the activities of architects and entrepreneurs go well beyond financial motivations, it's clear that many practitioners have sought and found enhanced financial compensation for the additional value they've delivered to their client, customer, or the market in general. Typically, such enhanced financial returns are a reflection of the additional risk they've embraced as active entrepreneurs. John Nash and Charles Bulfinch are two noteworthy examples of practitioners who have experienced the painful financial downside that comes with higher risk ventures that don't pan out. Underpinning all of these objectives is the reality that architects often hold values and a professional orientation that transcends their attachment to client service or a traditional form of practice. Often, such architects may not secure clients that share their objectives or a project in which to accomplish them. For these practitioners, enterprising architecture is a fundamental element of finding professional satisfaction in the purpose and nature of their work.

CONCLUSION

The examples from history and contemporary practice make clear both the viability and potential value that an enterprising architecture offers. The conventional practice of architecture as a consultant to a client will likely persist as the primary avenue for architects to contribute to the built environment. There is opportunity, however, for those in the profession committed to such a traditional form of practice to embrace a more extensive range of client interests, and incorporate practical issues that comprise the knowledge base in real estate as an integral part of the architect's competency. For other professionals, the traditional method of project delivery simply may not provide sufficient opportunities for them to make the contribution they intend. Diminished opportunity with clientele would likely have been exacerbated by the recent economic downturn. As the economy begins to recover over the next few years, some practitioners may find attractive alternatives by internalizing expertise and business activities in real estate. What's clear—from having studied and interviewed a number of practitioners—is that the profession lacks a rigorous framework for understanding and organizing the issues inherent in such hybrid organizations. Through the course of this study, clear and relevant models emerged that begin to give more definition to the organizational variations available to potentially integrated practices. These models provide a framework that allows architects interested in the integration of real estate to consider the implications of pursuing a broader range of associated opportunities.

A few significant points are worth noting in conclusion. First, external constraints such as the current economic uncertainty represent opportunities for entrepreneurial architects to reconsider their expertise and how it's applied. Architects over time have demonstrated the viability of fundamentally redefining their practice in the face of these and other external factors. However, a cautionary note is also warranted; making reactionary and uninformed changes in practice or project engagement in the middle of a financial crisis can be fraught with obvious risks, especially as one collaborates with partners who are facing similar financial challenges. Second, the profession at large stands to gain from integrating a basic level of real estate expertise, even if many are not prepared to take on real estate investment risk. Clearly the profession is a stakeholder in the real estate industry—indirectly sharing industry risk—and would be wise to leverage an expanded knowledge of real estate in making decisions regarding clients, project delivery, and design. Third, for those that do pursue a hybrid enterprise, a clear vision and definition of purpose is essential. At the core of these objectives is the architect's ability to initiate a project with certain aims, especially when those aims or values aren't shared by their traditional clientele. Fourth, there are significant costs and benefits associated with integration that must be considered even if many are difficult to quantify. Some of the significant issues that elude quantification are expanded organizational independence, effects on marketplace image, and achievement of .professional objectives, such as the goals mentioned above. Fifth, architects that engage in the real estate marketplace as speculators must recognize their relative strength in product differentiation compared to firms primarily operating a real estate enterprise. Competing on a low-cost basis does not appear to be a likely path to success.[30] Sixth, professionals must gauge their underlying loyalty to the traditional model of practice, given the peculiar conflicts that are present when orchestrating development projects in conjunction with design services for clients.

While it's difficult to ascertain the growth that may occur in these integrated forms of practice, opportunities remain for entrepreneurs to embrace the challenges associated with independently conceiving, designing, financing, and implementing their vision for the built environment. These opportunities are likely to increase for those committed to pursuing them over the next few years. The occasion provided by the current economic crisis, may be another opportune moment to carefully consider the potential of such divergent activities. Our present age of uncertainty has and will continue to shape the built environment in both predictable and yet unknown ways. Such uncertainty appears inextricably tied to our increasingly rapid pace of change. One can't isolate the impact of such change on our built environment from its collateral impact on the profession that is largely responsible for shaping it.

NOTES

1   Dana Cuff and John Wriedt eds., "Architecture: The Entrepreneurial Profession," in *Architecture from the Outside In: Selected essays by Robert Gutman* (New York: Princeton Architectural Press, 2010), 33.

2   Ibid., 32–42.

3   George Barnett Johnston, "Professional Practice: Can Professionalism be Taught in School," in *Architecture School: Three Centuries of Educating Architects in North America*, ed. Joan Ockman (Cambridge: MIT Press, 2012), 371.

4   Mark Linder, "Disciplinarity: Redefining Architecture's Limits and Identity," in *Architecture School: Three Centuries of Educating Architects in North America*, ed. Joan Ockman (Cambridge: MIT Press, 2012), 298.

5   Andrew Saint, *The Image of the Architect* (New Haven: Yale University Press, 1983), 6. Of additional relevance is Saint's vignette on entrepreneurship and the profession in the chapter, "The Architect as Entrepreneur."

6   Spiro Kostof, ed., *The Architect: Chapters in the History of the Profession* (New York: Oxford University Press, 1977), v.

7   H.H. Stevenson and J.C. Jarillo, "A Paradigm for Entrepreneurship: Entrepreneurial Management," *Strategic Management Journal*, no. 11 (1990): 17–27. Quoted in, Vesa P. Taatila, "Learning Entrepreneurship in Higher Education," *Education + Training*, 52 (1), 48–61.

8   H.H. Stevenson and David E. Gumpert. "The heart of entrepreneurship," *Harvard Business Review*, 63 (2) (March 1985): 85–94.

9   For a more thorough review, see Nathan Richardson, "Architecture & Enterprise: A History, Practice, and Analysis of Architectural Extensions into Real Estate" (master's thesis, Harvard University Graduate School of Design, 2009).

10   Howard Colvin, *A Biographical Dictionary of British Architects 1660–1840* (London: J. Murray 1954), 908–909.

11   Ibid., 48.

12   Andrew Saint, *Architect and Engineer: A Study in Sibling Rivalry* (New Haven: Yale University Press, 2007), 231–242.

13   For a more thorough review, see Nathan Richardson, "Architecture & Enterprise: A History, Practice, and Analysis of Architectural Extensions into Real Estate"

14   John Summerson, *The Life and Work of John Nash, Architect* (Cambridge: MIT Press, 1980), 9.

15   Charles A. Place, *Charles Bulfinch: Architect and Citizen* (New York: Da Capo Press, 1968), 56–93.

16   RIBAJ Editors, "Architect/Developer John Portman," *RIBA Journal*, 12 (December 1977), 504–509.

17   Among others, consider Wagner's Majolicahaus, Perret's 25 bis rue Franklin, and Barragán's Las Arboledas.

18   As an organizing framework for business in general, see Michael Porter, *Competitive Advantage: Creating and Sustaining Superior Performance* (New York: Free Press, 1985). Regarding integration across the value chain, see Michael Porter, *Competitive Strategy* (New York: Free Press, 1980).

19   American Institute of Architects, "The Business of Architecture: 2003 AIA Firm Survey," American Institute of Architects in partnership with McGraw-Hill Construction (2003).

20   Among those who hold a claim to specifically addressing environmental issues in their integrated structure of practice are Sir Clough Williams Ellis at Portmeirion and Randy Brown with Quantum Quality Real Estate at their Hidden Creek Development in Nebraska.

21   "The Architect as Developer," *Architectural Record*, 10 (Oct 1971), 126–127.

22   For instance, the Piano Craft Guild in Boston, Massachusetts. Leland Cott, interview with Nathan Richardson, 2009. See also http://www.brunercott.com

23   Avi Friedman, "Developing Skills for Architects of Speculative Housing," *Journal of Architectural Education*, 47(1) (Sep 1993), 49–52.

24   Chris Krager, interview with Nathan Richardson, Feb 12, 2009. See also http://www.krdb.com/.

25   Jonathan Portman and Jonathan Barnett, *The Architect as Developer*, (New York: McGraw-Hill, 1976).

26   Edward William Henry Jr., *Portman, Architect and Entrepreneur. The Opportunities, Advantages, and Disadvantages of His Design Development Process* (Ann Arbor, MI: University Microfilms, 1985), 373.

27   The American Institute of Architects, "2007 Code of Ethics and Professional Conduct," from the Office of the General Counsel. http://www.aia.org/aiaucmp/groups/aia/documents/pdf/aiap074121.pdf.

28   John Peterson quoted in: Katie Weeks, "Designers of the Year, John Peterson and John Cary, Public Architecture," *Contract Magazine*. January 2009; 50, 1.

29   Nathan Richardson, "Architecture & Enterprise: A History, Practice, and Analysis of Architectural Extensions into Real Estate"

30   For more on competitive strategy regarding cost, differentiation, and other factors, see Michael E. Porter, *Competitive Advantage: Creating and Sustaining Superior Performance*.

7

Certain Uncertainties: Architecture and Building Security in the 21st Century

David Monteyne

*[A]s we know, there are known knowns; there are things we know we
know. We also know there are known unknowns; that is to say we know
there are some things we do not know. But there are also unknown
unknowns—there are things we do not know we don't know.*
United States Secretary of Defense Donald Rumsfeld in 2002

This infamous epigram, uttered by Secretary Rumsfeld as his nation ramped up for
wars on terrorism and on Iraq, succinctly captures the spirit of today's "building
security" approach to architecture. Buildings, in this conceptualization, must be
designed to take into account certain uncertainties—even uncertain uncertainties.
We may be uncertain of the exact nature of threats to buildings and occupants,
but we can be certain that some of them will become victim to non-routine
events, whether natural or manmade. The nagging certainty of these uncertainties
forms the impetus for a discourse and practice of security among architects and
other building professionals, and a concomitant investment of money, effort, and
concern by building owners and managers, and occasionally occupants.

For architecture, every age is an age of uncertainty. While heat and cold, rain
and wind, can be predicted with some certainty for a particular building site,
extreme temperature swings, storms, and natural disasters have remained largely
unpredictable. We attempt to rationalize these non-routine events through
structural calculations, designing for fifty-year flood levels, or past seismic ratings.
That these and any architectural solutions, however certain, are proffered within
social, economic, and other contexts, results in their uncertain success. Unofficial
modifications to buildings, "grandfathered" urban fabric pre-dating building codes,
the everyday deviation of people from everyday life paths, social segregation and
environmental racism, individual and communal behavior—all may render even
predictable events unpredictable in their consequences. It is the uncertainties
arising from these contexts —often writ large, as geopolitics— that the discourse
and practice of building security increasingly seeks to address.

Building security professionals propose cost-effective and tested solutions to
perceived natural and humanmade threats against physical plant and human

assets. Meanwhile, their critics point to the broader costs to society incurred when security concerns trump "good design" in the public realm. This is, in fact, a debate that echoes from as far back as the 1950s at least, when architects and planners participated in civil defense planning against the threat of atomic attack. Then, as now, certain uncertainties prompted more and less subtle design interventions, and accompanying struggles over the characteristics and ethics of them. In the first decades of the 21st century, "security" broadly defined has been a powerful force producing the spatial experiences of people around the globe. Specialist architects and building security experts have provided the forms and details that have resulted in this securitized built environment. Except in the aftermath of 9/11, though, most contemporary architects have ignored the issue of security, sometimes to the detriment of their design intentions: their creations have been encrusted *ex post facto* with hectoring signage, surveillance cameras, and metal detectors.

This chapter discusses some of the implications of building security for the design and everyday experience of the public realm, mainly in the North American context. It avoids discussing buildings such as airports or purpose-built bunkers, for which security is a (often, *the*) chief programmatic feature. It also leaves aside the discussion of home security as a field outside the purview of most architects, although we might usefully understand the perceived threats to the home and family as intertwined with perceived threats to public buildings and their occupants.[1] In this sense, the fear of personal and property crimes operates in tandem with fear of terrorist attacks or other political acts in public space. The design solutions for these untoward events overlap, informing each other and materializing the sense that we live in an age of uncertainty.

A HISTORICAL PERSPECTIVE ON BUILDING SECURITY

Previous research I have conducted on United States architecture and civil defense during the early Cold War offers a historical perspective on professional design solutions to, and debates around, various perceived threats of the past sixty years. The threat of atomic attack first allowed architects to present themselves as experts in protective design or, as it was sometimes called, "design for survival." In a 1951 booklet, the American Institute of Architects (AIA) addressed its members, calling "attention to the desirability of their participation in the civil defense program so that full use of their professional talents and abilities may accrue to the greatest benefit of the public."[2] AIA committees, with names and mandates such as "National Defense," "Safety in Buildings," and of course, "Civil Defense," linked up with federal agencies tasked with preparing the nation for atomic attack and other perceived threats.[3]

From the start, the geopolitical threat was intertwined with the danger of natural disasters—as were the architectural solutions to these different problems. We might track the definition of the threat historically, as it shifted across overlapping moments since the opening of the Cold War. Beginning with news of Hiroshima, experts and ordinary folks struggled to understand the magnitude of destruction by drawing parallels with earthquakes and hurricanes. The initial threat of atomic explosions and their resulting firestorms gave way in the late 1950s to fallout radiation, when the victims of blast and heat were written off as impossible to save. As a radioactive dust, fallout in particular was tied to unpredictable weather patterns. Then, fallout shelter receded as the main goal of civil defense, and was replaced with protection from urban unrest in the later 1960s. By the early 1970s, the energy crisis redirected attention to threats and solutions derived from the natural environment (with some impetus from geopolitical negotiations), while prominent industrial accidents like Three Mile Island were reminders of humanmade excesses. The Reagan era saw a return to nuclear saber-rattling, as well as new articulations of a threatening urban underclass. Meanwhile, a series of U.S. embassy bombings raised the specter of terrorist acts targeting buildings, a danger that became domestic during the 1990s at both the World Trade Center and the federal building in Oklahoma City. The dramatic and tragic event of 9/11 brought these latter threats to a new prominence, though as this brief review suggests, a sense of threat to building security has been cultivated continuously since 1950.

At any given cultural moment, the definition and preeminence of the threat drives —but is also driven by— the formation of official and professional responses to it. For example, FEMA (now part of the Department of Homeland Security) is a direct administrative descendant of the Cold War civil defense agencies. AIA participation in Cold War civil defense planning paved the way for its current activities as adviser to government on a broad range of issues, building security among them. The early architects for civil defense also helped to establish the role of expert consultant or specialist on issues of building security; government-sponsored training as fallout shelter analysts permitted architects to develop and claim an area of scientific expertise supplementing their seemingly less rational function as designers. In competition with engineers and builders, architects labored to justify design commissions through a value-added approach—building security would be one of the "comprehensive services" that architects would offer to clients.[4]

The architectural solutions to security threats have not changed so much since the atomic era, or even before that. Whether designing protection from fallout, from urban rebellions, energy crises, the homeless, or a truck bomb, the security *parti* remains the same. Some combination of mass and distance, or what the early civil defense architects called barrier and geometric shielding, is necessary to prevent external forces from threatening or damaging valuable infrastructure and "human assets." Depending on the perceived threat, and on the site and context of a particular building, *mass* can be achieved through the choice of materials, the design of wall sections, reduced fenestration, earth berming, or other landscaping; *distance* can be attained through a setback from surrounding streets or buildings, a circuitous or layered entry sequence, programming functions in core or upper areas, the orientation of the building in relation to forecast sources of trouble. The ancient Mycenaeans, in their fortified hill towns, erected thick masonry walls that encircled inhabited areas, but also controlled the approach of enemies along a particular route: a combination of mass and distance. Likewise, Renaissance and early modern theorists of fortification deployed massive barrier walls and earthworks according to elaborate geometric strategies responding to the development of ordnance.[5] While we might think of building security solutions today as being more sophisticated or high-tech, the principles of mass and distance seem eternal.

BUILDING SECURITY TODAY

For a couple of years following 9/11, architecture journals unsurprisingly were replete with articles related to security from terrorism. In particular, many typological and morphological concerns were voiced about skyscrapers and central business districts, even though one of the targets hit that day was a low-slung, suburban office building (and, no doubt, a paragon of building security on the ground). These journal articles became increasingly scarce, partly due to the receding of perceived threats.[6] Moreover, the advice and examples scattered across the journals was assembled in three main design handbooks published in 2004–05, and sponsored by, or closely associated with, the professional associations. *Security Planning and Design: A Guide for Architects and Building Design Professionals*, was published by the AIA itself. *Security and Site Design: A Landscape Architectural Approach to Analysis, Assessment, and Design Implementation* was written by professionals closely associated with the American Society of Landscape Architects (ASLA). Finally, *Building Security: Handbook for Architectural Planning and Design*, a collection of chapters by a wide range of experts, was assembled and edited by architect Barbara Nadel. With Nadel's close connections to AIA headquarters, all three books bear the mien of official publications. They officially establish, as it were, security as a critical component of designing the built environment.

Briefly, several topics are common across the advice given in these books to architects and other designers responsible for building security. In different ways, building security design strives to control flows: of people, of vehicles, and of weapons effects. For example, the experts suggest that pedestrian *access* to buildings be limited to a single point; strategic lobby design can centralize many security features, but especially the screening of individuals attempting to enter the facility. In the case of an emergency, the people allowed in also must be evacuated quickly and safely; thus, *egress* was reconsidered carefully following the collapse of the World Trade Center towers. The principle of *progressive collapse* was illustrated manifestly on 9/11, but structural engineers were aware of its potential destructiveness as far back as World War II; structural redundancy and continuity are design solutions to the danger of upper floor loads exceeding the strength of a damaged structure lower down in the building. Another crucial issue raised by the explosive flow of forces unleashed by bombs is the material and assembly of the *building envelope*, especially fenestration. Since the Cold War, security experts have debated the feasibility of reducing the glazed area of building façades, a controversial proposal since occupants tend to like windows; today, laminated glazing and proprietary window films can mitigate the hazard of projectile shards, even in buildings with older kinds of glass. Meanwhile, mechanical engineers worry about the introduction of *chemical, biological, and radiological (CBR) agents* via building intakes, loading docks, or mail rooms connected to building supplies and returns; CBR filters, first developed during the early Cold War, are supplemented by strategies for the locating and protecting of mechanical systems.

Many of these threats to buildings and occupants can be controlled, according to the experts, by careful site planning. Comprehensive *perimeter security* comprises the first ring of what building security planners see as concentric layers or zones, each with "different security risks and responses."[7] The proximity of public streets, parking lanes, and sidewalks has become a crucial issue for building security professionals, due to vehicle-delivered explosives. Thus, designers aim to maximize *standoff distance* between a building and a potential point of detonation; anti-ram devices and other perimeter security components are complemented by strict controls over underground or adjacent parking areas.

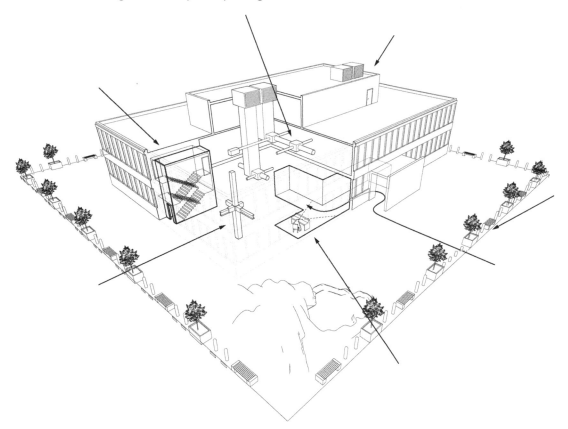

7.1   Beginning with arrow that curves and snakes in through the front entrance of the building; then proceeding clockwise.
1. Single, controlled point of ingress.
2. Site design elements, such as bollards, benches, and landscaping closely spaced with buried anchors, enforcing minimum setback of building from street.
3. Structural continuity to prevent progressive collapse.
4. Multiple points of egress for easy evacuation.
5. Air intakes with CBR filters, remote from intruders.
6. Reduced fenestration or laminated glass. Drawing by Shota Vashakmadze

Broadly defined, *building hardening* is the sum of these multiple interventions between the street and the building's core. "Hardening" is a Cold War term that initially had a specific meaning related to protection from the destructive blast, thermal, and electromagnetic effects of nuclear weapons. But by the late 1960s, the term had been adopted to refer to fallout protection, and then to describe securing a building against human aggressors on the ground. Building security professionals now talk about "target hardening" in the context of property crime prevention.[8] These design solutions make a facility harder to attack in the first place, and also render it more difficult to damage when attacked. According to the building security experts, hardening a facility also has a deterrent effect, such that perpetrators will search out softer targets. While this may be true in some cases, the deterrence factor was not in evidence in the choice of the Pentagon on 9/11, or of military installations targeted for other attacks.

All three handbooks attempt to promote a more or less calm and rational response to security design issues, in contrast to the widely lamented makeshift measures taken immediately after 9/11—for example, the jersey barriers set up around every public building. As the authors of *Security and Site Design* argue, these "[k]nee-jerk responses can actually increase the perception of threat, and instill fear, rather than promote a secure feeling."[9] The handbooks are in agreement that it is the perception of threat and security that is at issue for designers of the built environment. Fomenting fear can be avoided by "achieving transparent security," by which is meant security "not visible to the public eye."[10] In this demand for subtlety in security design, however, these experts seem to undermine the propagandistic effects of opaque or explicit building hardening. Security professionals are well-meaning, of course. Nadel states the overall goal for her collection as "balancing security and openness in a free society," a theme echoed in the other two design guides, Interestingly, the AIA-endorsed book also holds up as a high priority the balance between security and budget.[11] Building security, like any other supplementary expertise, becomes an element of professional promotion: it is the experts who know how to strike that balance between safety and freedom. Meanwhile, it has been a longstanding goal of the AIA to counter the cliché that architects are always going over budget.

It is precisely a sense of balance that is offered as the contribution of the design professions in these handbooks. They differ somewhat in how that balance is to be achieved. As the promotional body for the profession, the AIA is adamant that it is design, not hardware or electronics, that reveals the path to building security; that design provides long-term capital solutions rather than unpredictable, fleeting, and ongoing operations costs.[12] In Nadel's introduction to *Building Security*, though, she recognizes as a key concept that security professionals must "integrate design, technology, and operations." This valorization of an integrated approach reflects the breadth of the book, and its deployment of "over 100 multidisciplinary experts."[13] The most wide-ranging, detailed, and uneven of the books, *Building Security* covers everything from operations planning to property management, codes to professional liability for architects, and entries are written by facilities professionals, lawyers, and others. Several chapters focus on specific building types, such as arenas, religious buildings, abortion clinics, and courthouses. Chapters written by engineers introduce essential concepts of structural design and the protection of building systems. The book's breadth results in some inevitable inconsistencies in approach. For instance, the chapter on religious institutions, submitted by the Anti-Defamation League, argues that a most valuable effect of building hardening is to emphasize the appearance of security in order to deter attacks. Here, opacity rather than transparency, again becomes the key goal of building security interventions.[14]

In contrast, all three books aspire to reassure architects and critics who deplore the development of a fortress mentality out of the dictum of building security. The AIA book asserts outright that bunkers are not necessary, that "glass symbolizes openness in a democratic society" and can and should be used liberally in protective design.[15] This is essentially a carbon copy of the debate that raged around fallout shelter design during the early 1960s. Architects, critics, and glass manufacturers defended an open architecture of democracy —1950s glass-box modernism— against a tendency toward bunker architecture. Bunkerism was evidenced equally in the fallout shelter design programs promoted by the AIA and among the avant-garde proponents of Brutalism. Despite repeated assurances that fallout shelters need not be bunkers, it was difficult for people to perceive security without solidity.[16] Ironically, "achieving transparent security" today might be effective against actual crimes or attacks, but could increase the fear of them— that is, transparency might be ineffective against imagined or feared crimes or attacks. Alternatively, the opaque security of a bunker architecture might address both the reality and the perception of threats, but undermine the symbols and relationships of a free society.

Landscape architects can avoid these issues of building hardening and bunker aesthetics by bracketing off their scope. In *Security and Site Design*, the authors note that their focus is limited to the space between the building envelope and the curb. They prefer to call this area the building's "setback" rather than the more militaristic sounding "standoff distance." In this space, site design "can deter and/ or minimize the damage created by terrorist attacks." In fact, alterations to the site may be the only solution for existing buildings, where hardening would require hugely expensive retrofits. The authors assert that "beautiful" design will conceal the protective role of landscape elements, especially if "familiar site elements" are integrated together "in a seamless, transparent manner." Bollards and benches, trees, and concrete planters were common features shaping landscapes before 9/11 and Oklahoma City. Today, in order to deflect speeding vehicles, designers need to ensure that these landscape elements are properly spaced (maximum 4' gaps), and properly anchored through "substantial below-grade modifications" that preserve transparency.[17]

MARKETING BUILDING SECURITY

Given the long history of security design, it is remarkable how 9/11 is deployed extensively as a framing device for the principles of building security expounded in these books. In a rhetorical mode akin to ambulance-chasing, 9/11 delivers the urgency and horror necessary to convince architects and their clients of the handbooks' wisdom. Nadel's *Building Security* collection is the most explicit about making these connections. According to her Introduction, 9/11 and a number of "benchmark events" are foregrounded because one of the book's three key concepts is to "learn from the past." Now, as a historian, I fully support the idea of learning from the past, but one needs to be careful about what lessons are taken, and how one interprets them. There is an old saw that generals are always planning for the last war—is this true about building security professionals? For example, protective design inspired by the truck bombs used on U.S. embassies and the federal building in Oklahoma City, offered little to the security of buildings and occupants on 9/11.

We might fruitfully contrast the building security approaches advanced by the professional associations of designers, with those of the security industry (or, as some have called it, the "security-industrial complex"). The security industry addresses itself to building managers and others responsible for day-to-day operations of buildings or complexes. Professional architects and landscape architects are required to —indeed, they strive to— convince patrons to include security concerns at the design stage, thus adding to capital budgets that might be wrung from the stones of government or philanthropic sources, or the profit margins of speculative developers. Meanwhile, security industry consultants can play on continually updated fears to persuade building managers or landlords to pay for advice, master planning, surveillance systems, and armed patrols out of operating costs that can be passed on to tenants or written off as day-to-day business expenses.

The security-industrial complex often filters the language of threat through that of legality and liability. How much force can be used by private security firms to protect a building and its assets? What are the rights to public space surrounding a building? When should police be called in? And how liable are building owners or managers for damage to contents, occupants, and intruders? In conjunction with presumably temporary measures like jersey barriers, the continuing profusion of security camera systems, metal detectors, and guards are the most visible solutions proffered by the security-industrial complex. Less visible solutions include disaster plans and exercises, employee background checks, and the monitoring of personal communications. Crucial to the continued growth and profits of the security-industrial complex is maintaining threat in the foreground of discussions among those responsible for buildings. A weekly blog on www.buildings.com (the clearinghouse for information directed at "facilities professionals," and companion to the long-running *Buildings* magazine) keeps the threat fresh, and introduces the latest planning and technical solutions offered by the industry. Recent entries cover standoff distances, watching out for suspicious people taking photos or sketching a building (architecture students and historians, beware!), command centers, planning for corporate security, and new surveillance technologies. One article on public demonstrations warns of "emotional hordes" of non-specified protestors coming to "city streets across the nation."[18] Similar to the rhetoric of

Cold War "containment," which delineated domestic enemies within U.S. society, embattled building managers are made to seem beset on all sides.[19]

CRITICS OF BUILDING SECURITY

Critics of the security-industrial complex abound in everyday life, as ordinary folks regularly rail against the inconveniences of queues and searches, against surveillance and other invasions of privacy. Whether security measures increasingly instituted in public and private spaces actually ensure personal safety is a special consideration for skeptics who are convinced that these measures merely are meant to protect private property. Some take their critique public in the form of protests, performances, publications. For example, the political drama group Surveillance Camera Players performs skits for the cameras trained on sidewalks and other public spaces in New York and other cities. Their interventions in the securescape raise questions about citizens' performativity in public space, and about the audience for these performances. Do people modify their behavior for the cameras? Who is watching (if anyone) and how do they act on the information collected during surveillance activities? The Surveillance Camera Players, who view cameras in public spaces as contraventions of the right to privacy, also map their locations in order to both document the proliferation of cameras, and potentially, routes by which one could avoid them in a city.

A prominent voice among the critics has been the writer Mike Davis, who draws on William H. Whyte and others to ground his critique of security proliferation in analysis of the built environment. In his now-famous chapter on Los Angeles in the 1980s, Davis powerfully demonstrated the effects of fortress urbanism on the public realm of Los Angeles. He particularly targeted the bunker architecture of Frank Gehry as "a high-profile, low-tech approach that maximally foregrounds the security functions as motifs of the design." For Davis, this "'hardening' of the city" was the corollary of urban renewal.[20] Other critics have denounced the barricading of cities around the world, along with the gating of communities, and the "citadelization" of public space. Along with bolstered security forces and surveillance techniques, architectural and urban design elements allow for the efficient sorting of social groups, and contribute to the increasing privatization of public space.[21]

Architects and other designers also have been appalled by the hardening of buildings, and by the ugly security bric-a-brac layered over the built environment. Their criticisms often take a less political tangent than the Davis school, focusing on the aesthetics, experience, and function of buildings and public spaces. As we have seen, there is a lot of discussion within the professions about the need to strike a balance between building security and good design, the latter being some vaguely defined synthesis of commodity, firmness, and delight. These designers argue that security features need not be so apparent that buildings seem like ugly fortresses; that building security should not limit the public's ability to satisfy their practical, everyday needs; that public buildings ought to support, rather than suppress, the desire for human contact outside of the home and family.

However, striking a balance between security and something that architects would recognize as "good design" is fraught with contradictions. As Lawrence Vale has written recently, in his usual perceptive manner, "the tenets of the securescape work in precisely the opposite direction" from most of the best practices of architects and urban designers.[22] Denser neighborhoods and narrower streets; mixed-use buildings meeting the sidewalk; hiding cars away underneath buildings, or placing them next to the sidewalk in a "Main Street" configuration— all these current concepts for reviving public space directly contradict the building security guidelines emanating from the Department of Defense, General Services Administration, ASLA, and AIA. Vale also points out that contemporary building security doctrine can undermine the best practices of *past* architects, such as when the grand entry sequences of Beaux-Arts planning are circumvented, and citizens are diverted to securitized side entrances. Preservationists recognize the dilemma, but one wonders whether their arguments in favor of authentic historical experiences of architecture will prevail over the selling points of building security experts.

Finally, Vale notes the disingenuousness of much building security discourse. Researching Cold War architecture for civil defense, I also found that building security standards are technically arbitrary and socially unequal. As they did in the 1950s–60s, building security experts today make massive assumptions about the location of "ground zero," the magnitude of a bomb, the method of delivery, and so on. Any calculation of, say, standoff distance, is a result of these assumptions; but what if a bomber manages to stuff an extra 500lbs of explosive into a delivery van, packing 10–15 percent more punch? The extent and the characteristics of damage change. Building security experts have to decide on a set of numbers that then drive their guidelines and design practices. To give practical, real-world design advice, building security professionals have to set parameters that construct a quantifiable and calculable reality.

Moreover, a basic tenet of building security is that some buildings and occupants are more important than others. For example, GSA guidelines, adopted directly into the design handbooks discussed above, rank buildings by the likelihood of attack and criticality of operations.[23] Needless to say, DOD and other facilities supporting the national security state are considered highest-risk, requiring the strictest protection. Thus, military personnel, followed closely by high-level bureaucrats, receive the best building security, while ordinary civil servants warrant less. Meanwhile, there are no official guidelines for non-governmental buildings, such as the World Trade Center. Back in the Cold War, fallout protection at least was extended to all U.S. citizens through the National Fallout Shelter Program. Even then, however, certain top-level government personnel were provided significantly better *blast* protection in bunkers federally-funded through the Emergency Operating Centers program.[24] The building security discourse, now as then, reproduces and reinforces the inequalities of a real world framed by experts on retainer to clients worried about their own safety, and the defense of their property.

CONCLUSIONS

Historians are loathe to predict the future, but I can foresee little change in the character of the imperatives, solutions, and contradictions of building security. Surely, the exact nature of the threats, and the specific calculations of risk, will change, just as they have continued to do since the beginning of the Cold War. There have been and will continue to be peaks and troughs of public fear, driven by events and contexts. But the persistence and consistency of civil defense discourse and professional design practices has been remarkable. Building security is driven by cycles just like capitalist economies, and the two cycles may be related. When economies are in the black, there is more capital to protect, and more to protect it with. However, if the building security industry enters a recession, does the threat not remain the same? Are venture terrorists less likely to invest in attacks during slow economic times? The building security formula is to assess levels of threat, and then strike a balance between security and openness within a particular client's budget—but of course threat assessment and the required level of security solutions are both uncontrolled variables.

It is perhaps this seeming lack of control over variables that renders building security both lucrative and controversial. Liberal theorists might say that risk is an inherent aspect of modernity, globalization, and growth. Critical theorists counter that threats are (re)produced by a hegemonic state in the service of social control and business profits. The practices and effects of building security in the city, according to two critical geographers, are "sustained by the manufactured certainty of uncertainty in an endless American war on terror." For them, security "overrides conceptions of the 'good life' in cities."[25] This is no doubt a danger, as some areas of some cities have undergone significant spatial restructuring in the past decade or more. Add in restrictions to civil liberties and a burgeoning security-industrial complex, and the result is increasing socio-spatial securitization.

On the other hand, as Anthony Vidler concluded in his widely-read *New York Times* editorial, published just after 9/11, "terrorism alone will not decrease the importance of city centers for the public life of societies."[26] The histories of anti-colonial terrorism in London and Paris, he notes, have not emptied these cities' streets, neighborhoods, shops, and museums. New York's vitality will persist in spite of building security interventions in City Hall Park or the Financial District. But most North American cities are not New York: hopefully their public spaces will survive and thrive, providing social sustenance in ages of uncertainty.

BIBLIOGRAPHY

Appy, Christian G. (ed.), *Cold War Constructions: The Political Culture of United States Imperialism, 1945–1966* (Amherst: University of Massachusetts, 2000).

Chafe, William H. and Harvard Sitkoff (eds), *A History of Our Time: Readings on Postwar America* (5th edn., New York 1999).

Civil Defense: The Architect's Part, (Washington, DC: AIA, 1951).

Davis, Mike, *City of Quartz: Excavating the Future in Los Angeles* (New York: Vintage, 1992).

De Hart, Jane Sherron, 'Containment at Home: Gender, Sexuality, and National Identity in Cold War America,' in *Rethinking Cold War Culture*, Peter J. Kuznick and James Gilbert (eds), (Washington: Smithsonian Institution, 2001).

Demkin, Joseph A. (ed.), *Security Planning and Design: A Guide for Architects and Building Design Professionals* (Hoboken, N.J.: J. Wiley & Sons, 2004).

Dudley, Michael Quinn, 'Sprawl as Strategy: City Planners Face the Bomb' *Journal of Planning Education and Research*, 21/1 (September 2001): 52–63.

Ellin, Nan (ed.), *The Architecture of Fear* (New York: Princeton Architectural Press, 1997).

Farish, Matthew, 'Disaster and Decentralization: American Cities and the Cold War', *cultural geographies* 10/2 (2003): 125–148.

Fickes, Michael, '8 Ways to Protect Your Building during a Demonstration', accessed May 18, 2012, http://www.buildings.com/tabid/3334/ArticleID/13382/Default.aspx.

Fickes, Michael, 'Guess Who's Coming into Your Building', accessed May 18, 2012, http://www.buildings.com/tabid/3334/ArticleID/11089/Default.aspx.

Fickes, Michael, 'Protect Buildings from Vehicle Attacks', accessed May 18, 2–12, http://www.buildings.com/tabid/3334/ArticleID/13693/Default.aspx.

Graham, Steven (ed.), *Cities, War, and Terrorism: Towards an Urban Geopolitics* (Oxford: Blackwell, 2004).

Gray, Mitchell and Elvin Wyly, 'The Terror City Hypothesis', in *Violent Geographies: Fear, Terror, and Political Violence*, Derek Gregory and Allan Pred (eds), (New York: Routledge, 2007).

Hay, James, 'Designing Homes to be the First Line of Defense', *Cultural Studies*, 20/4 (2006): 349–377.

Hirst, Paul, *Space and Power: Politics, War and Architecture* (Cambridge, UK: Polity Press, 2005).

Hopper, Leonard J. and Martha J. Droge, *Security and Site Design: A Landscape Architectural Approach to Analysis, Assessment, and Design Implementation*. (Hoboken, N.J.: J. Wiley & Sons, 2005).

Jencks, Charles, 'Hetero-Architecture for the Heteropolis: The Los Angeles School,' in Nan Ellin, (ed.), *The Architecture of Fear* (New York: Princeton Architectural Press, 1997).

Katz, Cindi, 'Banal Terrorism: Spatial Fetishism and Everyday Insecurity,' in Derek Gregory and Allan Pred (ed.), *Violent Geographies: Fear, Terror, and Political Violence* (New York: Routledge, 2006).

Low, Setha, 'The Fortification of Residential Neighborhoods and the New Emotions of Home', Special Issue, M. Van de Land and L. Reinders (eds) *Housing, Theory and Society*, 25/1: 47–65.

Mallory, Keith and Arvid Ottar, *Architecture of Aggression: A History of Military Architecture in North West Europe 1900–1945* (London, UK: Architectural Press 1973).

Marcuse, Peter, 'Urban form and globalization after September 11th: the view from New York' *International Journal of Urban and Regional Research* 26/3 (September 2002), 596–606.

Monteyne, David, *Fallout Shelter: Designing for Civil Defense in the Cold War*. (Minneapolis: University of Minnesota Press, 2011).

Morton, Jennie, 'Corporate Security Plan Tips', accessed May 18, 2012, http://www.buildings.com/tabid/3334/ArticleID/12966/Default.aspx.

Nadel, Barbara (ed.), *Building Security: Handbook for Architectural Planning and Design* (New York: McGraw-Hill, 2004).

Nasr, Joe 'Planning Histories, Urban Futures, and the World Trade Center Attack' *Journal of Planning History* August 2/3 (2003), 195–211.

Sorkin, Michael (ed.), *Variations on a Theme Park: The New American City and the End of Public Space* (New York: Hill & Wang, 1992).

Tyler May, Elaine, *Homeward Bound: American Families in the Cold War Era* (Rev. edn. New York: Basic, 1999).

Vale, Lawrence J., 'Securing Public Space', *Places* 17/3 (2005): 38–42.

Vidler, Anthony, 'A City Transformed: Designing "Defensible Space"', *Grey Room* 7 (Spring 2002): 82–85.

NOTES

1   See, for instance, James Hay, 'Designing Homes to be the First Line of Defense', *Cultural Studies*, 20/4 (2006): pp. 349–377; Cindi Katz, 'Banal Terrorism: Spatial Fetishism and Everyday Insecurity', in *Violent Geographies: Fear, Terror, and Political Violence*, Derek Gregory and Allan Pred (eds) (Routledge, 2006): pp. 349–61; Setha Low, 'The Fortification of Residential Neighborhoods and the New Emotions of Home', Special Issue, M. Van de Land and L. Reinders (eds) *Housing, Theory and Society*, 25/1: pp. 47–65.

2   *Civil Defense: The Architect's Part*, (Washington, DC: AIA, 1951), 7.

3   David Monteyne, *Fallout Shelter Designing for Civil Defense in the Cold War*. (University of Minnesota Press, 2011): pp. 107–113; Michael Quinn Dudley, "Sprawl as Strategy: City Planners Face the Bomb", *Journal of Planning Education and Research*, 21/1 (September 2001): pp. 52–63; Matthew Farish, 'Disaster and Decentralization: American Cities and the Cold War', *Cultural Geographies*, 10/2 (2003): pp. 125–148.

4   Monteyne, *Fallout Shelter*, pp. 48–50, 137–138.

5   Paul Hirst, *Space and Power: Politics, War and Architecture* (Cambridge, UK, 2005); Keith Mallory and Arvid Ottar, *Architecture of Aggression: A History of Military Architecture in North West Europe 1900–1945* (London, 1973).

6   Avery Architectural Index keyword searches related to "building security" reveal the pattern noted here. See also Joe Nasr 'Planning Histories, Urban Futures, and the World Trade Center Attack', *Journal of Planning History*, August 2/3 (2003): pp. 195–211; and Peter Marcuse, 'Urban form and globalization after September 11th: the view from New York', *International Journal of Urban and Regional Research*, 26/3 (September 2002): pp. 596–606.

7   National Capital Planning Commission, "General Security Design Solutions", quoted in Hopper and Droge, *Security and Site Design: A Landscape Architectural Approach to Analysis, Assessment, and Design Implementation* (Hoboken, N.J., 2005): p. 12.

8   Demkin, Joseph A. (ed.), *Security Planning and Design: A Guide for Architects and Building Design Professionals* (Hoboken, N.J., 2004): p. 45. See also Monteyne, *Fallout Shelter*, 159, for a discussion of the term "hardening."

9   Hopper and Droge, *Security and Site Design*, 8.

10   Barbara Nadel (ed.), *Building Security: Handbook for Architectural Planning and Design* (New York, 2004): pp. 1.8.

11   Nadel, *Building Security*, 1.5; Demkin, *Security Planning and Design*, p. 12.

12   Demkin, *Security Planning and Design*, p. 14.

13   Nadel, *Building Security*, p. 1.5.

14   Nadel, *Building Security*, pp. 17.15–17.16.

15   Demkin, *Security Planning and Design*, p. 16.

16   Monteyne, *Fallout Shelter*, pp. 160–167, 188.

17   Hopper and Droge, *Security and Site Design*, 18, pp. 26–27, 55–58, 172–174.

18   Michael Fickes, '8 Ways to Protect Your Building During a Demonstration', 12/28/2011. See also, for example, Fickes, 'Protect Buildings from Vehicle Attacks', 03/06/2012; Fickes, 'Guess Who's Coming into Your Building', 11/29/2010; and Jennie Morton, 'Corporate Security Plan Tips', 10/01/2011; all on www.buildings.com, accessed, 18 May 2012.

19   The era of "containment" has been seen as crucial to the reformation of concepts and practices of gender, citizenship, labor politics, and many other aspects of society. Historians trace the concept to George Kennan's 'Long Telegram.' George Kennan, 'The Necessity for Containment' (1946), excerpted in William H. Chafe and Harvard Sitkoff (eds), *A History of Our Time: Readings on Postwar America* (5th edn., New York 1999): pp. 14–19. See, for example, Elaine Tyler May, *Homeward Bound: American Families in the Cold War Era* (Rev. edn. New York, 1999); Christian G. Appy, ed., *Cold War Constructions: The Political Culture of United States Imperialism, 1945–1966* (University of Massachusetts, 2000); Jane Sherron De Hart, 'Containment at Home: Gender, Sexuality, and National Identity in Cold War America,' in *Rethinking Cold War Culture*, Peter J. Kuznick and James Gilbert (eds), (Washington, 2001): pp. 124–155.

20   Mike Davis, *City of Quartz: Excavating the Future in Los Angeles* (New York, 1992): quotations on p. 232, 240. For a more sympathetic review of this approach to defensive architecture, see Charles Jencks, 'Hetero-Architecture for the Heteropolis: The Los Angeles School,' in Nan Ellin (ed.), *The Architecture of Fear* (New York, 1997): pp. 217–225.

21   Peter Marcuse, "Urban Form and Globalization." See also the essays in Michael Sorkin (ed.), *Variations on a Theme Park: The New American City and the End of Public Space* (Hill & Wang, 1992), Nan Ellin (ed.), *The Architecture of Fear* (Princeton Architectural Press, 1997), and in Steven Graham (ed.), *Cities, War, and Terrorism: Towards an Urban Geopolitics* (Blackwell, 2004).

22   Lawrence J. Vale, 'Securing Public Space', *Places* 17/3 (2005): p. 38.

23   Demkin, *Security Planning and Design*, pp. 4–7; Hopper and Droge, *Security and Site Design*, pp. 11–15.

24   Monteyne, *Fallout Shelter*, pp. 211–225.

25   Mitchell Gray and Elvin Wyly, 'The Terror City Hypothesis', in *Violent Geographies*, Gregory and Allan (eds): p. 330 and p. 343.

26   Anthony Vidler, 'Aftermath; A City Transformed: Designing "Defensible Space"', *New York Times*, 23 September, 2001, reprinted as 'A City Transformed: Designing "Defensible Space"', *Grey Room* 7 (Spring 2002), p. 84.

8

The Uncertainty of Authority

David Yocum

THE PHONE CALL

Late on a Friday afternoon this past August, my business partner and fellow architect took a telephone call. A client of ours was on the line, with some unsettling news. The project in question was a private religious facility we had been working on over the last 2 years, navigating it through fundraising and design, and now the middle of construction. The Client had decided, unilaterally, to remove an important component of the project (let's call it a "*Significant Design Element*"). The Client was feeling the pressures of an extraordinarily tight construction budget and wanted to spend the potential savings somewhere else in project. My partner listened and then calmly protested. The client insisted. My partner further explained that the element was too critical to the project to delete, had previously been approved, and was about to undergo fabrication. The client became deeply unsettled, stating in effect, "How can you tell me what should or should not be in this project? This is not your building – it's ours!" The *Significant Design Element* would be removed.

4 weeks later, following heated conference calls, accusatory emails, latent fears of a contractual dispute—and an agreement between my partner and I that we were willing to be terminated over the matter—the Client retrenched. To their deep and vocal disappointment, the *Significant Design Element* remained in the project. In their eyes, on this point, we had failed them. In ours, the project had prevailed and rightly so. But what had put us collectively in this position and why?

THE PROBLEM OF AUTHORITY

Years ago, I was an instructor in an architecture "career discovery" program. My students were high school- and college-aged, and uniformly bright and eager. Having never taught before in my chosen field—for I myself was but a first year graduate student—I had little to offer beyond my own inquisitiveness. Having assembled my students in a circle for the first day of class, I blithely asked, "So, why do you want to be an architect?" The responses were straightforward, endearing, and already painfully familiar.

> *"My grandfather was an engineer, and I've always been good at drawing. I'm detail oriented. My mother said that I should think about a career in architecture."*

> *"I grew up in a suburban town that was especially ugly; I want to make the world a better place."*

> *"I've always been interested in the design of buildings. I've been taking courses in it for a while and am pretty talented."*

> *"I need to see my ideas built."*

There is little mistaking the desire which fuels the ambition of the aspiring architect, and it comes in many flavours. For the purist, it is a febrile curiosity about the physical world still unquenched: a need to investigate, to create, and to make anew. For the cosseted or simply confident, it is an entitlement to impart an aesthetic will into the world, and attain a titled career while so doing. For the altruist, it is a vision of correction, or betterment (for "our future", for "the environment") and the personal credo that it must be realized. The value of each approach is - like most ambition - morally relative.

The road to the practice of architecture is a long, arduous, and expensive one, requiring five to eight years of higher education and a professional diploma (now, most commonly, a Master's Degree). After that, typically, the same number of years again required in internship and practical training. Finally, a battery of certifying exams (7 at last count) are necessary to achieve the bloody victory of Professional Licensure. The young designer (now likely over the age of 30, indebted, and bone-tired) is then, legally, an *Architect*, with a capital A. They may now practice, so we might say, with confidence and authority.

Let's linger for a minute on that last word—authority. Conceptually, but also practically, we can understand authority as legitimate power. For example, lest we risk imprisonment, we respect it in the context of Government (authority *de jure—concerning law*). Police have authority. Legislatures have authority. Judges have authority. As citizens of the state, we (voluntarily or not) vest authority in our government and obey rules accordingly. By way of a second example, lest we risk our job and our income, we respect authority in the context of the marketplace of employment (authority *de facto—in practice but not necessarily ordained by law*). An employer has granted us a job to perform, and should we fail to respect the authority that comes with the terms of employment, the job will be taken away.

However, authority is not just vested in power. Authority can also simply be expertise in the absence of actual power. We talk of an expert being "an authority," a person with such extensive experience and knowledge that what they assert within their field of practice is unquestioned. Authority here is granted *socially* by the audience. We might tentatively call this "an authority of experience." At this point, given the evidence of expertise, the burden of proof is not upon the authority figure; the burden of counter-proof is upon the rest of us.

Accordingly, authority as we understand, practice, and respect it, carries both power and expertise. It invokes government, establishes social hierarchy, and, in the context of this discussion, underpins professional disciplines.

The practice of architecture is a licensed profession in every statewide jurisdiction of this writer's nation (the United States). It is upheld by rigorous standards of education, training, and testing, and is only conferred by legitimate regulatory bodies. One might then reasonably assume that the status of *Architect* carries real authority on all principal matters related to the practice. But it does not.

Let us make two lists. List 1:

- The layout of buildings (organization and function)
- The look of buildings (design, appearance)
- Materials, finishes, products arranged and deployed in buildings (contents and experience)

These are the activities that have traditionally defined the practice of architecture; they are the reason most aspiring architects have traditionally chosen the field. With some latitude, we can classify all of the above as *the things that architects wish to do, and enjoy doing.* However, with this desire comes no promise of authority. Setting aside minimum requirements regarding the design and construction of buildings - which are regulated by law, and which we will discuss below - once guided thorough education and training, the Architect practices effectively through the use of personal judgment. Yes, *personal judgment.*

Assuming each occupant can properly exit the building, a hallway can be 5 feet wide, or 6 1/2. Assuming there is at least a minimum amount of natural daylight in an office, a window could be round, or square. Assuming materials won't burn too quickly, window shading could be bright red sheer fabric, or dark mahogany mini-blinds. In each of these choices, which is the better decision? And who decides?

List 2 comprises those aspects of the practice for which architects have trained, have been licensed, and therefore carry an obligation to abide:

- federal laws, including accessibility (rights)
- state laws, including building and life safety codes (construction and occupancy)
- local jurisdictional requirements, including zoning (use and compatibility)

The above are often referred to as the "health, safety, and welfare" aspects to the practice. The architect, acting under a license granted by a jurisdiction, is the implementer of the authority of others. Put bluntly, the above are *the things that architects must do.*

So if we can again recall the varied attributes that induce the young architect to progress to the point of professional practice—curiosity, creativity, confidence—is easy to see how the promise of professional practice would be *seemingly* fulfilled.

With adequate skill, experience, and validation, the architect should be able to confidently practice across the full continuum of professional responsibility. However, in that there is a essential difference between the things architects seemingly *like to do*, and those that the architect *must do*, so too is there a fundamental divide between the authority the architect purports to carry, and the authority which the architect actually carries.

For all the activities of the first list that define—for wont of a better phrase—the "art of the practice", the architect is only an authority to the degree that others may agree. Expertise cannot be reasonably arrogated to oneself, it must be granted by a community of peers. While the community of practicing peers is long-standing, substantial, and supportive (academic, professional, social), the fundamental issue with the professional practice of architecture is that peers are almost never clients. The enabling authority in almost every case is a community of one—the Client.

THE CLIENT

A client is one who is in a position of need who retains the professional services of another. Attorneys, advertising executives, interior decorators, therapists, and architects offer services to clients. Clients have a need that they themselves cannot fulfill, and these needs are met by someone else who has the authority and experience to offer these specialized services. It is understood, through a relationship of trust, that if services are not satisfactory, they may be revised, or corrected. It could be thought that a *client* is therefore a *customer*, but this is not necessarily so.

When we discuss the transfer of the goods of services we must make a careful distinction between a client and a customer. The term *client* stems from Latin *cluere* (to hear), *cluens* (heeding), and cliens (follower, retainer). Associated etymological meanings include "to listen, follow, or obey" and "to lean upon". Historical usage of the term and concept is from the "client – patron" institution of ancient Roman class structure. Clients in need of services *leaned upon* those who were in a position of greater authority to *grant* services. Clients needed the protection of someone more powerful, and therefore they *obeyed* or *followed* the informed guidance of a patron, in exchange for services, money, or privileges in return. Whether or not money changed hands, patrons retained authority.

Quite different is the term *customer*—simply "a person who buys." A customer purchases goods in exchange for funds or items of equal value. There is a transactional simplicity in the seller-to-buyer relationship, and control is shared. We say "the customer is always right" because the customer can simply return the goods for refund. This is not necessarily the case with professional services, where, it can be argued that authority is retained by the one providing services, not the one purchasing services. Ponder this for the moment - *authority is retained by the one providing services*. How strange! In the contemporary marketplace of professional services, this historical perspective is unfamiliar, seemingly inverted, and perhaps a source of the problem.

THE DISCUSSION

Returning for a moment to our opening anecdote of the "Significant Design Element" [SDE], what transpired between the original telephone call demanding the change and the final decision to retract the threatened deletion? In one way, it can read as a detailed screenplay for this particular issue; in another sense it is not unlike disputes about aspects of design on any project where positions of authority are unclear.

Architect: *Thank you for your telephone call. While we appreciate your perspective, we cannot support such a deletion. The Significant Design Element is more important than other additions to the project you are considering. As well, the direction you are giving us goes against previous approvals.*

Client Leadership: *We disagree. Do as we say. Delete the SDE and proceed.*

Architect: *We appreciate the financial pressures and your preferences. However, what you are recommending is not in the best interest of the project. Alternatively, we have found savings elsewhere which are in line with the objectives of the project, and equal to the proposed savings of the SDE.*

Client Leadership: *What? There are things you can delete, and you didn't delete them earlier? Why are you spending this money? Why are you withholding this information from us? We instruct you to delete the SDE as well as the additional items you suggested, for double the savings. Proceed as directed.*

Architect: *We do not recommend deleting the SDE without our consent. We recommend you take the proposed savings we have identified elsewhere.*

Client Leadership: *How is this possible? It is our building! You cannot tell us what will be included and what will be deleted. Delete the both the SDE and the additional items you propose.*

Architect: *We are in midst of construction. It is not your building until the construction is completed. A change to the project requires the approval of the Owner, the Architect, and the Contractor. The deletion of the SDE is not in the best interest of the project. Again, we recommend you take the savings we have found elsewhere and proceed.*

Client Leadership: *Is that a threat?*

Architect: *In no way is it a threat, but we reserve the right to refuse your request. We have to. In doing what we think is best, we are doing our job.*

Client Leadership: *(Silence).*

Architect: *Please understand that (a.) it will cost you more in design fees and delays to the project schedule than if you simply leave the SDE in the project, and (b.) given the time that has now passed, the Contractor has informed us that it will increase the construction cost to delete the SDE.*

Client Leadership: *We will take the alternate savings you have proposed, but we resent this conclusion. You "screwed us."*

[4 months pass. The project is completed.]

Client Leadership: *Thank you for this building. It is incredible. Amazing. We love it. We appreciate everything you have done for this project.*

So what prevailed, and why? And who was correct? Consider the options.

- We, the Architects, prevailed! We had successfully convinced the Client on the merits of our advice! Not so. The decision had hinged not on our authoritative "recommendation" in the matter. It was made on timing alone, under the pressure of a potential contractual dispute. In the eyes of the Client, we had forced their hand: we had demanded they purchase something they no longer valued, because it was too late. Even more damaging, again in the eyes of the Client, by acting against their short-term will, we had held our own interests above theirs— the ultimate professional betrayal. While we would argue strenuously to the contrary, and for sound reasons, the perception of our actions trumped our motives.
- The Client was right! They achieved the savings they had originally intended! Again, not so. While the Client did achieve savings, it was not for the deletion they had desired, and the issue was no longer monetary. With the *leverage* of money, and for the limited *purpose* of saving it, in our eyes they had forced a confrontation that violated both the fundamental trust of the relationship and the stated priorities of the project. While the goal of frugality might have seemed noble in the short term, by seeking and achieving it, from our standpoint they had threatened the stability of the project, and offended our leadership in the process.
- The Project? Perhaps. Aside from the deep bruising, the project continued, to a successful outcome. While mutual trust and respect had been damaged (at least momentarily), each party achieved a key priority. From the Client's point of view, they saved the value of the money they had intended, even though the target of their deletion remained. From our standpoint, a key component of the design was intact, for the betterment of the project as a whole.

IT'S COMPLICATED

The relationship between Client and Architect is built on trust, but the specifics of this trust are unclear and uncertain. From the standpoint of the Client, every decision made by the Architect during the course of a project *should be* fiduciary. The Client might believe something along the lines of "the Architect acts at all times in the best interest of the Client, giving quality professional service, addressing the needs of the Client, and deferring to the opinions and choices of the Client." From the standpoint of the Architect, this is not necessarily the case. The Architect has a legal obligation to ensure that building codes, life safety, zoning requirements, and the like, are met, for sure. It is also expected that the Architect will draw and specify a building that will function appropriately. In the practice, this is referred to as a reasonable "standard of care." Beyond this, what basis is there for making decisions regarding the design and construction of a building? Is the Client's financial or aesthetic welfare more important than the welfare of the building? Should the more or less expensive option be taken? Should the higher or lesser grade of craftsmanship be required? Should the colors on the walls be of the choosing of the Client or the Architect? Who is the client for the building after the Client has sold the building? Who best determines the appropriateness of a building that stands for 100 years, the one *in need of expertise*, or the one *offering expertise*?

Herein lays a fundamental uncertainty of contemporary practice. In the absence of specific authority relating to all matters of the Architect's informed judgment, when the Client expects professional fiduciary service that is at odds with the Architect's advice, what preserves the Architect's authority? A community of peers? A good contract? The ability to walk away?

This author would argue that the root of the uncertainty rests with two issues. First is the difference between a client and a customer, and the shift of power from the *authority* of the Architect to the *ownership* of the Client. The second is the question of ownership in the first place, and how it intersects with the value of the project itself. Or, in other words, who owns the project anyway? Why is it important that one party has control over the other? And what determines if it is any good?

From a historical and culture perspective (if we think of the Roman *client-patron* structure), we understand that a Client comes to an Architect with a set of needs. Because they *need* to, the Client is expected to heed the professional guidance of the Architect. However, in the absence of any kind of social or political structure which preserves the Architect's authority on matters other than those already dictated at federal, state, or local levels, the Client perhaps believes that they are simply "purchasing services". The Client perhaps believes that they are simply a *customer*, as if the retainage of an Architect is as simple as a merchant transaction, whereupon if the Client is displeased with the services of the Architect, they can return the services rendered to date for a refund and choose a different solution.

One root of this uncertainty, discussed above, is that the duties of the Architect span both matters *de jure* and matters *de facto*. Actions relating to the former are indisputable; actions relating to the latter are entirely mootable. Said another way, all buildings must have red fire pulls, but not all buildings must be painted fire engine red.

The second root of this uncertainty is that Clients are almost always Owners, and this is a critical distinction. Although financial control of a project does not carry actual authority over architectural matters, it provides indisputable leverage. Owners are still Clients who are seeking knowledge and service that they themselves do not have and cannot perform. But with the ability to control not only the expenditures of the project, but also—through the control of construction budgets—what is built, and—through the control of fees—the solvency of an Architect's practice, they are in effect *purchasing* authority.

And so Owners, by retaining financial control over a project, have no need to heed matters *de facto* as they are offered by the Architect. In an endless marketplace of offerings and preferences, Owners need Architects, but not their opinions. And thus, we find ourselves in a foggy landscape where control is open to debate. The Client needs the professional services of the Architect, but does not want to be obligated to heed the guidance given beyond what is absolutely necessary or desirable. Likewise, the Architect is willing to render professional services, but not in a relationship where their expertise will not be heeded. The result is a relationship where authority is both uncertain and negotiable.

THE AUTHORITY OF PRINCIPLE

And this is the rub. Architects operate in their very own profession with limited control. Beyond the laws that they are required to implement, the essence of their practice is conducted with an illusion of authority. Clients want to trust in the judgment of their Architect, but often do so only when it suits them.

This happens at the start of projects, when prospects are bright and few decisions have been made, and often decreases exponentially through the course of the work. By the time a project is complete, the Client who questions the authority of their Architect is common, if only because by the conclusion of the project, the Client has come to know the project well enough to assert his or herself on any number of aspects of the final outcome. After all, when Clients are Owners, they— quite naturally—perceive that they retain the right to choose.

Similarly, Clients trust in the judgment of their Architect easily when funding is adequate, but decreasingly when the cost of the work becomes realized. And when a project costs more than a Client is willing or able to pay, the Architect's guidance is only validated when it is financially palatable. Why pay more for something simply because the Architect says it is important?

The reason Architects have little authority is because no Client is bound to agree with matters of an Architect's *individual* judgment. There is no personal or legal imperative to do so. The only time a Client absolutely must agree is when the course of action is otherwise bound by entities that carry authority *de jure*.

This is what separates the profession of architecture from peer professions. Imagine you, the reader, want to avoid a penalty for reckless driving; should you not choose to follow the guidance of your attorney, you risk legal penalty. Imagine you have contracted pneumonia and wish to be healed; should you not choose to follow the trained expertise of your doctor, you risk dying. Imagine you wish to commission a building (a house, an office tower, a school); as long as you accord to the strictures of zoning, building codes, and the like, nothing the Architect insists on matters all too much. Inelegant and awkwardly proportioned houses sell in numbers greater than elegant, well-proportioned ones. There is a market for anonymous office blocks just as robust as for idiosyncratic expressive towers. School buildings that resemble warehouses more than spaces of delight and learning are ubiquitous. In the end, a Client's choice to save money and build something ugly might draw the ire of the neighborhood aesthetes, but the Client will suffer only to the degree that he or she is willing to suffer socially.

If we return a last time to the anecdote of the "Significant Design Element"—we see precisely where the illusion of authority is most present in practice. In a situation where authority was uncertain—the inclusion of a particular design element—the decision to *withhold* control set up a serious conflict. The Client wanted control and the Architect would not cede it. Even though the Client could be faulted for reversing a previous decision, it would appear that the Architect was breaking the fiduciary bond that should have been upheld, in the eyes of the Client. This author would argue to the contrary. The prudence of going at odds with a Client depends entirely on the significance of the issue and the principle of the stance. In the case of the Significant Design Element, the Architect deemed the element important enough that its removal would irreparably damage the project, and that alone was enough to thwart the will of the Client.

This author would caution Architects against believing that they are in control of a project in any way beyond those responsibilities for which they have legitimate authority. To assert authority under illegitimate pretenses (personal judgment, relative aesthetic opinion, taste) sets up an unstable power relationship that is easily toppled. Architects do this all the time, vehemently arguing for a certain design outcome, believing that they have authority in the matter, and hoping that it is recognized as such. [*If I can just convince my client that I am right!*] This may work for a while, but only until the Client decides otherwise, especially when they are an Owner. All a Client needs to say is "Forget it, I don't like it", and stand firm, and the veiled persona of the authoritative Architect is stripped bare.

As such, Architects would be better advised *to avoid the assumption of authority entirely*, and understand that all projects are a result of conflicts of interest. The priorities of the Architect are not necessarily those of the Client, or the Owner if they are one in the same, or the Users, or the Public at large. The underlying principles of every project are negotiable, and the degree to which the principles are negotiated sets about the value of the work in our culture.

In this sense, we have to question the authority of ownership on the part of the Client. The case is an easy one to make when the project is for the public. Courthouses, schools, parks, and the like, belong to a collective group, and therefore offer the opportunity and the responsibility to base decisions on a set of shared principles. The case is harder—but but equally critical one to make—when the project is for a private individual, where priorities are often defined by personal preferences, or even prejudices. The challenge for Architects is to separate the notion that financial or legal ownership carries any authority beyond face value. Buildings regularly outlast individual terms of ownership; they expend far more natural resources than originally conceived; they exist in the world as collective assets and liabilities.

Architects have to be acutely aware of when a Client is making a poor decision relative to the underlying integrity of the project. There is no reason for an

Architect to ascribe their practice to same standards as a Client. More frequently than would be expected, especially under financial duress, decisions made by a Client can irreparably damage a project. It is at this moment that other tools of leverage are advised to be invoked (i.e. intrapersonal, financial, or contractual) to assert the significance of an issue. In the Significant Design Element anecdote above, my partner and I advised the Client that they would be in breach of the contract if they removed an element during construction that was against the approval of the Architect. While this was a correct interpretation of the contract in hand, it was an extremely narrow and risky reading that carried the danger of a substantial contractual dispute, even termination. And this is precisely the point. In a relationship in which authority is uncertain, the position of the weaker hand is often the strongest. Authority does not rest on the basis of power, but rather on the integrity of the principle being asserted. This is often called "safeguarding the best interests of the project." Certainly, Clients retain financial and other controls. Often, they negotiate better leveraged contracts. However, when the integrity of a project is placed above a Client's own immediate desires, the willingness of the Architect to sacrifice his or her own involvement serves as a surprising affirmation to the merits of an argument.

Make no mistake about it; Architects have a reputation for being knaves, and accurately so. They are often seen to serve up design that is merely personal, or prevailing, preference, and thinly veiled. In the absence of clear priorities that can be commonly valued, and an unwillingness to get into the business of negotiating the underlying worth of projects, Architects remain both marginalized and frustrated. And so a funny thing happens when instead an Architect asserts—beyond all of the things that a project *must* be—not what they necessarily *want*, but what they believe to be *right* and *why*. And what is *right* for a project is not necessarily the same thing as what a Client wants, or thinks is right. Or, for that matter, when that Client is also an Owner, what that Owner necessarily wants to pay for. Suddenly, all the more quickly, it becomes clear what is important, and Clients have the right to choose to agree or not. Projects succeed or fail; practices thrive or dissolve. Perhaps our shared built world is filled with one less enfeebled building.

While no meaningful projects are accomplished without leadership—"You can't design a building by committee," remains a tiring but truthful adage—what we find is that the uncertainty of authority lays the groundwork for the realization of the underlying value of a project. When Clients challenge the perceived authority of Architects, and Architects challenge the assumed ownership of the Client, the value of the exchange, and the outcome, is more rigorously vetted. Motivations may have to be questioned, assumptions might be abandoned, and egos are likely to be bruised, but the worth of the thing is cultivated. Some might label this "partnering" or "collaboration" or "teamwork", but this author would argue that no such weak tonic cures the underlying ailment. It is rather the struggle for principle, and the extent to which it is negotiated, directly equates to the value of the work, and the degree to which it is achieved.

Postscript: For sure, in any project of merit, budgets are limited, priorities shift, and relationships can be messy. However, the client for the project on which this article is based, Congregation Or Hadash Synagogue, in Atlanta, Georgia, remains to this day one of our most valued and respected. While the uncomfortable particulars of the story are true, they are shared here to make a professional argument rather than a desire to share the details of a personal one, and remain but a very short chapter in an overall process that was mutually respectful, supportive, and deeply gratifying. The members of the Congregation remain not just our clients, but our friends.

Upstream Imagination

Jennifer Bonner and Christine Haven Canabou

In a land far away, in a rarified brand of disaster environment, fellow members of the undeterred dreamt to see the light of day. It had been one long downpour of studio criticism. This is 2008, before the summer began. It was Protagonist who dusted off a copy of an old encyclopedia, flipping, obviously, to *Natural Sciences, Plate 17, Figure. 43, Kopp's Volumeter*.[1] Only half of the measuring device was found in the engraving, an inky delicacy of line weight and authoritative precision. So, with the utmost gravity, she scanned and then mirrored the image to construct a belly cavity from it. Then, she turned to *Plate 76* and appropriated *Pinna rudis* for what would become an apparatus of drinking hoses. Finally, in this order, the efficient creature paged to *Plate 20*, filled with engravings of electric and magnetic instruments. "I located what looked to me like a spring," she explained. "This became the creature's chance for mobility in the form of pogo stick." With nearly-perfect, upright posture, the hydration-filtration animal was fabricated. Concerned voices gladly chimed in, other peer reviewers stared blank and despaired. In one hand, Protagonist clung to her trusty illustrated tome of engravings of natural history and other oddities. In the other hand, she held a collection of dreams from dreamers.[2] The doubt lodged deep in the critics' throats arrived in synchronized form: "Protagonist, do you really know what you're doing?"

No response. She closed her eyes. Protagonist took herself outside the stuffy room, to a morning when the air was very fresh. Imagine a river, she thought, any river. Stretch it for miles, like soft boardwalk taffy, and then snake a course. Here, specifics on place, distance and geography don't matter. Particular hydrological and scientific readings, washed away.[3] What matters is that the water moves. A little. A lot. Up. Down. Swells. Shrinks. It nourishes and connects, undercuts and divides. Water, acting out gravity, flows and branches. Water bends around time as the event. It marks time. If time is narrowed and short-circuited, like the meander bends of the Mississippi, claimed by man-made miracle systems such as levees and dams, *all in the name of progress*, then, in due course, history tells us that nature will come to reclaim its version of energetic progress.[4] It's just a matter of time before a flood is unleashed.

This is the story of a rising river, its improbable menagerie and an odd lot of history's floating objects, reconceived. (Fig. 9.1) Frankly, the journey ahead and its

9.1   Fictitious
river

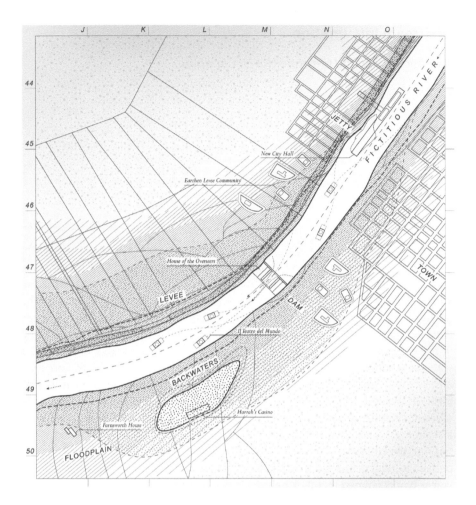

bends, breaks, and assorted baggage are entirely invented.[5] At the moment, the Collector of Curiosities awaits, downstream.

5'-3" OFF THE GROUND

> flood·plain n. 1: level land that may be submerged by floodwaters 2: a plain built up by stream deposition[6]

The sun is breaking on the River, and Collector of Curiosities, a watchful creature of flight, is performing his morning shuffle of junk under the mighty Farnsworth House. (Fig. 9.2) Completed in 1951, the twentieth-century icon, in all of its glass and steel tranquility, is located on a *floodplain*. The site has flooded more than 60 times. In photographs, though, man's masterpiece is presented as romantic and removed from its untamed environment. In many ways, the house simply ignores its paradoxical setting.

But the paradox hardly eludes the house's new owner. A hoarder who happens to have wings (metallic scales and flesh-colored feathers), Collector has renovated the masterpiece with the latest in flood-resistant terrazzo flooring. Large panes of glass communicate a "wash-and-dry" architecture—one that finally accepts and accommodates decades of floodwaters.[7] *His* Farnsworth House reveals new types of systems for architecture and urbanism along a watercourse.

Along the Fictitious River, the Gissen-brand of mud, weeds, dankness, and other subnatures is more than an unpleasant site condition to be put in its nature box.[8] These inconvenient weeds are more than the stuff of outside, in which the limited tonal range of the inside draws inspiration and views. They are more than inconvenient, viewed through the framed precision of an impeccably controlled inside.

In Collector's mind, the flood (and muck and debris) is just one of life's facts. This morning is a letup from the rain, but the shores are already beginning to darken along the swollen River. Daily, he senses the perfunctory threat of rising water. Some of the other creatures affectionately say that Collector has giant-magnifying glass eyes, because he notices everything—from washed up flotsam, such as church pews, parts of Ford Mustangs, and broken-down exercise machines to bloated financial documents, underwater mortgages, and runaway healthcare premiums. To collect his daily prize possessions, or survey the land for what he calls "junk data," Collector relentlessly combs the River. (In the wake of the global economic crisis, the times require it.) Upon return, he stuffs and archives his finds under the house.

Ludwig Mies van der Rohe, who was likely concerned with potential flood levels, as well as an aesthetic statement, designed the house to float 5-foot-3-inches off the ground. In the Miesian world, the house was neither lifted high enough to fully escape the ground nor allow occupation of the space underneath. Collector, squat, by human standards, at four feet, doesn't differentiate between the house's interior and its exterior underbelly. In the Miesian diagram, the topic of landscape and seasonality trails off into a tidy binary between nature and space. That framework is as seductive as a kid-leather Barcelona chair: Slide on in, gaze off at the picturesque landscape—out there, *removed*—and escape life's weeds.

By constrast, Collector's Farnsworth responds to the inherently time-based quality of nature. When water rises, just wait after the morning-lull, four feet of murky water will eliminate the terrace (one of the horizontal planes) from the figure of the project. Anchored but afloat, the Miesian Glass Box, what the late Philip Johnson referred to as a "pure" and "undisturbed cage," will come to absorb the entirety of its surroundings.[9]

Consider the stagnant water, humidity, and condensation. The old "cage" hid from the realities. Part of a system of watercourse urbanism, the new cage invites them, opportunistically, through a reimagined building envelope that's far more permeable. (In France, there was talk of a double-skin wall that provides space for insects![10]) Here, the Farnsworth is re-rendered semi-opaque through a similar strategy for the environment. However, the bugs and brown water are not confined to a glass cavity, as if a slice of sealed, swirling ant-farm diorama. When at last the waters will recede, as they do, and steamy dampness pervades, Collector can live in a protective, thickened quilt of breathable air and still connect with the outside world.

Surprisingly, it's in moments of floodwater crisis and its sticky aftermath that his bandwidth is the broadest. That's when he maximizes connectivity between the River he serves and the ecology. Collector, who tweets, blogs and chats, has no concept of privacy. He's what's called open-source. Thus, they say he steals from the creatures living along the River; he ruffles some bright, flashy feathers, pops out, God Bless, a pop-up, and then shares code. Not obfuscated code! He not only watches the fluctuating waters and its ripple effects, as they ebb and flow, change over time, self-organize, encounter its most fragile moments, but also fortifies and amplifies those very same principles of the River's ecology through obsessive documentation—on and through the house itself. We're told the documentation is a catalyst for transformation: What changes? What endures? What is fickle? As it turns out, the sectional, indoor-outdoor quality to all of Collector's physical-digital junkyard collecting has infused the reconceived home with a profound sense of time; water, its fluctuations, marks it. Collector's downstream curiosities over the years have been hoarded and then washed away. Today, for instance, all that's left are digital posts and scrapbooks, washed back on the house in endless video feeds. Collector and the Farnsworth demonstrate one way for a home and its owner to stay put in flood-prone areas and accept the swells of the river.

ACCESS DENIED

> back·wa·ter n. 1.a: water backed up in its course by an obstruction, an opposing current, or the tide b: a body of water (as an inlet or a tributary) that is out of the main current of a larger body 2: an isolated or backward place or condition[11]

Late morning, farther upstream: Slot machines, artificial lights, and surveillance cameras combined with disorientating carpets and concealed pontoons constitute the large, thickened gambling floor of Harrah's Casino. Welcome to man-cave civilization. Suddenly, as if on cue, the skies open up, leaves shiver.[12] Built in 1996 by a developer, the casino floats inconspicuously over *backwaters* of the Fictitious River. Housed within this 150,000-square foot container, the casino, in all of its alien monumentality and proud-casino typological wrappings, predictably ignores the waterfront.[13] Yet, the casino depends on—and is bound to—the water as an alibi for operating under local regulatory law. The casino also doubles as the headquarters for Alarming Sow, an efficient but reluctant broadcaster of all things bad. (Fig. 9.3)

Today, it's *bad*. The casino is already underwater in millions of dollars of debt. Amidst the continuous casino hum—ticking, chirping, and ding-dinging of the slot machines—Sow issues even higher decibels for a double whammy: debt and natural-disaster warnings. Zig-zagging back and forth the big, dumb gaming container, Sow requires good peripheral vision. Most crisis most of the time seems to happen overnight. But crisis rarely *just* happens. There are warnings—slow, simmering warning—that build up and intensify out of direct view long before an eruption. The warning signs might even be in *direct* view. At the stubborn casino, the gambling parties on "through hurricane, blizzards, and national crisis" remaining open "every minute of every day."[14] All the while, debt and hazard risks rise.

Mother Nature does not like debt. Abundance of abundance? *That can't last.* Sow sounds the alarm. Code pink escalates to the highest level—plum.

The rain pounds, the river surges and suddenly, all at once, water tops the banks and rushes to dry land. On a normal, devastation-free day, the Fictitious River flows at a rate of 450,000 cubic feet per second.[15] At that typical flow rate, it would take about four minutes to completely fill one of the world's largest buildings. Today, on flood day, it takes less than one minute. The flood rate is some 2 million cubic feet per second. With that kind of volume, the flood could put a foot of water over the state of Louisiana in just under a week.[16]

Years from now, the casino will become a full-blown green-house with a more responsive relationship to the water's edge. Today, though, tops of half-submerged trees bob up, tracing the perimeter of tomorrow's gambling floor. In the future, gambling will still abound. But there will be an agile, layered framework behind the economic exchanges and social transactions that the old casino took to dangerous

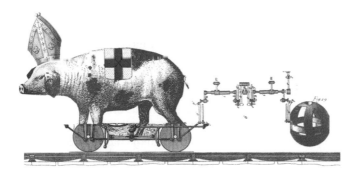

9.3   Alarming
Sow

excess. The casino-greenhouse, while primarily a money-making enterprise, will calibrate and recalibrate exchanges, within a larger system, to spur growth and renewal—to grow, in time, through measured results, checks and balances *with* the overall ecosystem. New trees and plants along the edge will respond to the water's temperament. The new casino-greenhouse and its landscaped watery edge will move beyond a singularly one-dimensional framework of instrumentality: What's in it for me![17]

On the much-needed ecological renovation, Alarming Sow reports: "The artificiality and complexity of an alarmingly large space releases function from its defensive armor to allow a kind of liquefaction." Sow pronounces the last word in slow, hushed fashion, as if a secret, as "li-que-vacation." Continues Sow: "Programmatic elements react with each other to create new events. The alarmingly large space returns to a model of programmatic alchemy."[18] The future, according to Sow, seems mighty bright: Gaming and ecological environments blur—*cha-ching*!

Still, the persistence of the past, and its excess, can't be washed away. In fact, it is marked by water, like the Farnsworth, on the casino's perimeter. Today, swelling water levels and the greater system's triple bottom-line is also registered on the rows of slot machines, and they, in turn, inform the external environment.

BUOYANCY AND BOARDED UP 56′

> barge n. 1.a: any of various boats: as a: a roomy usually flat-bottomed boat used chiefly for the transport of goods on inland waterways and usually propelled by towing; b: a large motorboat supplied to the flag officer of a flagship; c: a roomy pleasure boat; especially: a boat elegantly furnished and decorated[19]

Against the powerful current, the journey continues. As the late-afternoon rain intensifies, the waters grow thicker and muddier. Precautionary bordering on neurotic, Flood Instigator, a usual suspect of the nature-event's unleashing, holes up behind the boards, behind his protective shell, waiting. (Fig. 9.4) The "boards," as Flood Instigator calls them, make up the exterior of an 82 foot tall structure that was built by Aldo Rossi in 1979. Situated on a barge-like platform on the Fictitious River, Il Teatro del Mondo isn't a Venetian lacey floating pavilion from the sixteenth century. The building, which houses a theatre, looks more like Jan Sadeler's depiction of Noah's Ark—a vessel to withstand the *flood-of-all-times*.

This mighty ark not only suggests resilience in the midst of catastrophe but also the possibility of interior refuge. Rossi flanked the entire exterior of the steel scaffold structure with a homogeneous application of wood, as if the blank façade were a boarded-up storefront. That's eight stories of civic program. Closed for business. Indefinitely.

Instigator takes off his hairshirt.[20] (This flood is brutal.) In bad times, the thorny creature unplugs the swords and blades from his armored shell, cues up movies such as *The Clock*, and takes a bath.[21] He slides into rising, soapy water, which is contained in porcelain in a room on the building's highest occupiable level at just over 56 feet. Floating, he's frothy-light in warm escapism. Below, on the Fictitious

9.4   Flood
Instigator

River, creatures and objects attain, as if cruel whimsy, varying levels of buoyancy. Similar to the closed-shop theatre aesthetic, when disaster calls—and it's a callin'— homeowners board up windows, doorways, and thresholds during the onset of warnings for catastrophic weather.

As night falls and flood levels continue to rise, the disaster weary and the River ecology become more coordinated and calm. But *no*, not Instigator! The creature causes great hardship by thrashing all about in the River, overflowing the levee, and repeatedly surging the banks. The shelled loner, longing for friends, is thought to be aggressive. But he's actually just a part of the larger system, dutifully responding to and giving spatial qualities to the reams of data feeds about the mood of the River's environment. Some days, like today, he can be especially agitated. The tub is where he washes away anxiety and steel-blade mustiness from his soft body. But the bath tub scum-ring persists.

In the morning, after the storm, the boarded-up theatre will be towed from place-to-place via tugboat. (It's clear that it can't continue to stay in harm's way.) The promise of Rossi's float implies that housing and other entertainment programs might also fluctuate within high floodwaters rather than remain fixed to the conventions of the cul-de-sac or the repeatedly flood-prone Main Street. Traditional building typologies are static. Conventional land-based housing can't relocate when mild flooding occurs. Instead, living rooms, kitchens, and bedrooms become soggy fixtures in standing water. Water in motion—fast or hesitant— reveals architecture immobile—a poetic device architects have long played with, from buildings to entire cities such as Venice.[22] But in times of crisis, unwavering rootedness may be the surest way to sink. In the boarded-up theatre project, architecture is mediated by the River, with time in tow. Escapist-active engagement for all!

14'-8" SPEW

dam n. 1: a body of water confined by a barrier 2: a barrier preventing the flow of water or of loose solid materials (as soil or snow); especially: a barrier built across a watercourse for impounding water[23]

The stars are out, tiny dancing blurs on the black oil-slick water, and just ahead, past the sharply swerving current, there appears to be a French-style house spanning the entire waterway. Claude-Nicolas Ledoux, an alleged water-fiend, imagined a body of work in which buildings occupy water and inhabit embankments. Up ahead, lodged between a narrowing of the River, voilà—the House of the Overseers of the River (1779). It resembles a dam. The project straddles two topographic elevations—high, basically, stepping down to low. Some creatures marvel at the project's engineering feat; other creatures say it's all about the pretty waterfall. At the moment, *we* just wonder how to pass this strange, stubborn barrier.

Up ahead: Spewing water. The house does not defer to the flow of water as part of a natural gravity-fed system. The house spews. Or perhaps, water spews the house. The material is projectile shot from the interior. It's all quite conspicuously interactive. There's a dynamic, if linear, fluidity between inside and outside, coursed through time, *if the River is time*, channeled through a fantastically scaled environment.

The home itself combines the volume of a countryside mansion (retreat), and infrastructural gate (greet). As usual, *hate to intrude*, but the only way to reach the upper river basin is to turn down our motor and navigate through the dark, cold living room, to the freight elevator under the water feature. A sign on the door signals that the structure is fine, as is the homeowner. Rain Maker, an intensely perky one who pogos on a mechanical leg to collect river water, emerged unscathed by the flood. (Fig. 9.5) Now, in calmer waters, against the comfortably steady crash of the house's falling water, she appears practically blissful. The home not only displays the labors of the River but also filters, renews, and replenishes them—all conveniently in one location. Long ago, Ledoux replaced the home's central hearth with thousands of gallons of grayish water. Now, water bisects the house like the

9.5   Rain Maker

Panama Canal. There's a nearly constant flow of water that plows through the living room. The bad water from one end shoots through the house and comes out clean on the other end. That's the spew. On the end in need of filtration, buoys bob up and down in the River, and Rain Maker attaches one of her many rubbery macaroni hoses to the filtration system to help the cause. The hoses also help with major drainage issues.

Like the radicalized interior that she occupies, Rain Maker, whose spherical belly contains, filters, and redistributes the River's water, is a holistic, flexible response and adaptation to her environment. In this regard, Rain Maker, an unapologetic student of *Ecological Urbanism*, finds herself at home.[24] The net-like skin covering her belly has built-in elasticity—expanding and dilating, with time, and the River's fluctuating capacity. For Rain Maker, the performative "spew" redefines the terms of her environment. There is no tidy koi pond sequestered to the backyard, or obligatory hearth, stage center. There's this: exuberance, atmosphere, conduit, and adaptive filtration all wrapped into one. Out of the very rise and fall of the river, the House of the Overseers emerges hopeful among ghost river towns and ruins.[25] It demonstrates that anticipatory, responsive design can deal with disaster of all sorts—even the more quiet, everyday, intractable kind.

WATER'S EDGE

> *lev·ee n. 1. an embankment for preventing flooding 2. a continuous dike or ridge (as of earth) for confining the irrigation areas of land to be flooded*[26]

Upriver, just before sun-up and around the bend, views open up to the Earthen Levees (2010), an attractive, do-it-yourselfer of mound constructions that ring a community of single-family homes. The community suffered considerable damage, but that's expected given the age of the mounds. Luckily, Levee Builder is back at at work. (Fig. 9.6) The enthusiastic, perhaps too, some say, earth-moving creature provides immeasurable services. Much larger than most elephants, Levee Builder, who goes by Lavie, a polymath whose books range from *Landform Building* and *Monsterpieces* to all things umwelt theory and foodie instructional, is always on her toes.[27] With big boots, hopelessly muddy, her body is a machine for lifting. She uses her trunk and large pyramidal-shaped legs to level the ground. She also brings water and carries materials to the site. During off hours, her soft underbelly, supported by toy-Toblerone legs, offers a natural source of shelter from life's harsher elements—including *this* bendy ride.

How did we arrive here? Back in the hushed confusion before the crisis, when weather agencies warned of unprecedented flood levels, ponderous dons such as Alarming Sow maintained radio silence, hoping that would quell fears (and Mother Nature's harsher side) until there was a need to "really" sound the alarm. Meantime, in response, the ring of mud "banks" popped up at a rapid clip. Then, came a cheery focus—a preoccupation, really—on the levee surface. And then, perhaps out of nerves, or notions of desperate survival *displaced*, that surface just got bigger and smoother. Imagine: Bunches of trunks, deployed, to smooth one

9.6   Levee Builder

continuous surface; some folks fantasized that might make the crisis go away! The real mania, though, probably came when a magical Grasshopper sauntered on to the site. Suddenly, the elephants as earth movers could press their massive pads into sticky ground and *architect*. Brilliantly, how it was, and they, it seemed, that the quasi-scientific sculptural mud formations happened just so effortlessly, as if they plugged their prehistoric sixth toe into a smart ground.

Crisis is a good time, too, to talk about resilience. The mounded community represents latent potential of bounce-back topographies, but it also raises questions about constraints. How to bounce back so that it really drives us forward, not backwards, down the road? In many parts of the world, there are deep cultural and historical associations to land ownership. In the U.S., the single-family levee has the potential to re-appropriate protective barriers into a system of meaningful topographies. In that effort, though, it also potentially seriously disrupts, even exacerbates, the problem it's trying to solve: flooding. Interestingly, levee landforms alternate between remarkably different conversations, from *Landscape Urbanism* (a totally conscious exchange within the discipline) to *Junkyard Wars* (unconscious conversation outside of the discipline).[28] During a bad patch, though, it's the latter, more mundane stuff that often becomes a more compelling source for actually producing work. For houses situated in the countryside, the fence is generally used to keep out livestock and demarcate property lines. For houses located in suburban communities, and for those in much closer proximity to one another, the fence is used for privacy and to forewarn others of security measures. During times of flooding, the fence, once a thin wooden construction, is reconfigured into a thick earthen barrier that can reach six- to ten-feet in height. In planometric view, if the fence transforms from a thin line to a deep poche, it begins to work as a landform linked to a topographic system. Protective, integrated into the land, the wall construction encircles entire territories with a series of thick rings similar to those used to fortify medieval cities. With bodies as ditch-digging machines and backhoe equipment, Lavie and her team sculpt the yard into an engineering marvel, or at least feel as if they've made mountains, or mounds, move. And in many ways, they have. Here, the process of making is as regenerative and reaffirming as the physical construction. But the push and pull with the land is a bargain with nature and its own elaborate balancing act.[29] If floods or other nature disasters have taught us anything, it's that nature has the last word.

SUBMERGED, HALF WAY UNDER

jetty n. 1.a: a structure extended into a sea, lake, or river to influence the current or tide or to protect a harbor b: a protecting frame of a pier 2: a landing wharf[30]

Sliding through still, fatigued waters and soupy late-morning heat, we arrive at City Hall. It's unforgettable. Yet, commissioned by a trade magazine, Samuel Mockbee and Coleman Coker's New City Hall (1999) was published once and apparently forgotten.[31] On an island along the Fictitious River, however, the structure is the happy home of Remora, the anxious but disciplined creature made up of a limp bouquet of suction cups with special sensors. (Fig. 9.7) Rivaling a thirty-story building, the city hall climatic monument is identifiable from the existing urban terrain and re-imagines urbanism in the context of disaster. It is 1,400 feet long, an offset of the River's edge.

Look closely, and things are *not what they seem*. In plan, the offset is a first clue. In section, the entire structure is anchored to the riverbed. If there's a thought that we've just encountered another version of Rossi's floating barge project, then the clues quietly dismantle it: The narrow civic "bar" functions as a *jetty*. During an event as extreme as a 100-year flood, when vulnerability abounds, the city hall

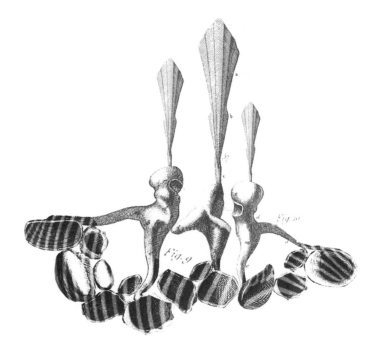

9.7 Remora

functions ideally.[32] Stationary and unable to float, the structure is offered up—as if a sacrificial act. On one level, it is inundated with floodwaters. On another level, it protects the shoreline by redirecting the River's currents. Remora, whose name, in Latin, means to "delay," equivocates, with discernment, between and among; she suspends, floats, accommodates. Underwater, she measures and records the environment's activity through mesh sensors; her smooth cups, navy with cobalt mesh, form momentary attachments to essentially ask, How's everything? The city hall's underwater section is largely infrastructural. Comprised of turbines, thermal exchanges, and mechanical systems, the civic institution can also be a responsive barrier. Time is built into its resilient framework. Designed to perform in fluctuating states, the City Hall as *jetty* defends the nearby urban grid.

Remora's mark of progress is largely invisible. Initially, it's the kind of quiet, behind-the-scenes, underwater, infrastructural work that beguiles until it's aired and hung out to dry. The jetty, as figure, is a stationary object embedded in a dynamic, fluid ecological system; the Mockbee-Coker project deploys an animated-form that both registers and demonstrates this variability. Creatures like Remora inform and transform the object and the greater system; as does time; as do fluctuations in seasons, water levels, and nature's bounty. Here, the large public ground acknowledges the fluctuations of waters levels—and anticipates festivals and other happy communal activity when it's not overtaken by floodwaters. Half of the building's programmatic mass, including the mayor's office, is submerged. The result is a free plan in the form of a public promenade, which is slightly above the River's waterline. Watching from above—who else?—Remora. When she can live on land, Remora does; her suction cups fan into a large non-structural sail rising above the public ground to announce climatic and ecological nuances of the River. The data has been tracked from her time submerged. Delicate and billowing, the canopy visualizes the imperceptible river currents while also responding to wind patterns. The colors rise and fall, from frothy whites to deep blues.

Traditionally, city hall, and all of its authoritative trappings, is positioned in the city's core. Not at the waterfront. Especially not offshore, in the cold, dark depths of some river. Civic monumentality demands stepping up to higher ground! Here, in Remora's world, the sequential order of engaging the building is flipped. Did Mockbee and Coker have her in mind? Rather than climbing up to access the building's ground, pedestrians descend from the River's embankment by way of an underwater tunnel. The gradual descent is more readily reminiscent of entering a subway than a civic building. Conventionally, infrastructure is located underground. But on the Fictitious River, city hall defies easy categorization and anticipates broader interpretations of public ground.

SYNTHESIS

"Really, we're waiting," the bewildered critic says, sighing. "Do you really know what you're doing?" Bewildered, too, Protagonist pauses, wondering if this is just a trick question, and then asks a question: What is to be gained by taking a monster-creature and having it occupy one of man's most esteemed architectural icons situated in a dreamt-up river community of other design marvels from other historical moments with other monster-creatures and have the water-course system, suddenly, all at once, flood? Design, at its most hopeful and productive, is an anticipatory discipline. It helps give form to future ways of living and working. In discontinuous and difficult times, more than ever, what's needed is a dynamic look at the future of situations, rather than objects alone or species type. In extreme circumstances, most people most of the time play it safe. But what if we were to be inspired by what we know and where we could go instead being paralyzed by fear?[33] Extreme circumstances also present an opportunity, to a point, for extreme learning—compressed, fresh, clear-eyed thinking. In this state, heightened awareness is extended from body to behavior, from home to environment. Greater anticipatory modes of design could be a natural extension. Here, the journey catalogs a make-believe collection of future behaviors instigated, enabled, and animated by something as elemental as water—taken to excess. Times may be dogged, but it's precisely these times when basic imagination takes on extraordinary power. A crisis, it's been said, is a terrible thing to waste. [34]

NOTES

1   J.G. Heck, *The Complete Encyclopedia of Illustration* (New York: Park Lane, 1979): 38–41, 95. Originally published in 1851 as *Iconographic Encyclopedia of Science, Literature, and Art*. A series of plates were pilfered and re-appropriated.

2   Jorge Luis Borges, *The Book of Imaginary Beings* (USA: Penguin Group, 2006). Originally published in *El libro de los seres imaginarios,* (Buenos Aires, 1967). Here, Borges curates "strange creatures conceived down through history by the human imagination." The menagerie to follow borrows from dreamers as diverse as they are impressive, including Confucius, Shakespeare, Kafka, Poe, and C.S. Lewis.

3   For the remainder of this essay, the intention is to negate hyper-contextual readings of place and time. To "un-learn," we've learned by now, is rethinking preconceived ways of knowing, doing, thinking, being. Of moving through the world. Here, it's about how one comes to know the Farnsworth House, Il Teatro del Mundo, Harrah's Casino, etc. A map of Campo Marzio by G.B. Piranesi (1762) disregarded the literal city of Rome and in turn collapsed reality with fiction. If Piranesi sets up a formal project for the city, the Fictitious River sets up an ecological project mediated by water.

4   See Mark Twain, *Life on the Mississippi River* (1883. New York: Random House, 2007): 3–4. "It seems safe to say that it is also the crookedest river in the world, since one part of its journey it uses up one thousand three hundred miles to cover the same ground that the crow would fly over in six hundred and seventy-five." The prominent "crookedness" indicates that the river is in a state of constant fluctuation and "thus straightening and shortening itself" over time. Curious Protagonist's preferred river of inquiry is the Mississippi. Specifically, the mighty Mississippi River captivates because of its rich history, politics, nuances, and power to carry great misfortune.

5   On the notion of lists, see Umberto Eco, *The Infinity of Lists* (New York: Rizzoli, 2009): 118. Eco makes a case for "lists" and their cultural relevance by describing the distinction between practical, pragmatic, and poetic lists in reference to catalogues, visuals, mirabilia, collections, treasures, and general excess. An excerpt regarding

the "poetic list" further equates invention with lists found in literature: "It is obvious why people make practical lists. But why do they make poetic ones? In part, we have already explained this: because we cannot manage to enumerate something that eludes our capacity for control and denomination, and this would be the case with Homer's catalogue of ships… Homer was not interested in knowing and telling us who the leaders of the Greeks really were. He, like the bards who came before him, was inventing. This would not make his list less referential except the fact that instead of referring to objects in the real world, it would refer to the objects of his epic world…"

6   *Merriam Webster's Collegiate Dictionary*, Tenth Edition, 1993, 447.

7   Tim Love and Elizabeth Christoforetti, "Wet 'n' Dry City" in *Architecture Boston*, Winter 2013 (Volume 16 n4): 42–43.

8   David Gissen, *Subnature: Architecture's Other Environments*, (New York: Princeton Architectural Press, 2009).

9   Farnsworth House on the Internet 2012. A National Trust Historic Site. 12 January 2012 <http://www.farnsworthhouse.org/history.htm>.

10   Gissen, *Subnature*, 174–79. See R&Sie(n) *Mosquito Bottleneck* project.

11   *Webster's*, 85.

12   "Leaves shiver" when it starts to get real ugly out. See Mark Twain, *Adventures of Huckleberry Finn*, (1885. New York: Dover Publications, 1994): 95.

13   Rodolfo Machado, *Monolithic Architecture*, (Munich: Prestel, 1995).

14   Michael Sokolove, "Busted: Can the Model for Casino Gambling be Fixed?" in *The New York Times Magazine*, 18 March 2012, 39.

15   As a reference, the Mississippi River's flow rate varies from annual averages of over 700,000 cubic foot per second (cu ft/s) to around 200,000 cu ft/s. It depends on the time of year, with flow rates highest in the spring and lowest in the fall.

16   Calculations extrapolated from assumptions and references in James Parkerson Kemper's, *Rebellious River*, (Boston: Humphries, 1949): 7.

17   Mark T. Conard, *The Philosophy of Martin Scorsese*, (Lexington: University Press of Kentucky, 2007).

18   Rem Koolhaas, "Bigness: The Problem of Large" in *Wiederhall* 17 (1994): 32–33.

19   *Webster's*, 92.

20   See biblical reference for "hairshirt". Flood Instigator wears a hairshirt to induce pain (a kind of obedient repentance for causing devastation throughout the land).

21   Daniel Zalewski, "The Hours: How Christian Marclay created the ultimate digital mosaic," *The New Yorker*, 12 March 2012, 50–63. See Christian's Marclay's film, *The Clock*, which opened at the 54th Venice Biennale 2011 and subsequently won the Golden Lion award. Also see, Rosalind E. Krauss, "Clock Time," *October*, Spring 2011, No. 136, 217. Krauss hailed Marclay's transformation of "the reel time of film into the real time of waiting." Many of the creatures in this essay pose similar questions about "the real time of waiting" and the architectural implications of temporal, cyclical waters as well as the "appropriation" of past legacies.

22   Juhani Pallasmaa, *The Embodied Image: Imagination and Imagery in Architecture*, (United Kingdom: John Wiley & Sons Ltd., 2011) 50. Encountering water and stone, such as in great cities such as Rome, has been said to be metaphysical. Water's "hesitancy" animates what is perceived to be the most enduring, most steadfast

aspects of architecture. On water's presence poeticizing architecture, the author cites the work of Carlo Scarpa and Luis Barragán. And the whole town of Venice.

23   *Webster's*, 291.

24   See *Ecological Urbanism*, edited by Mohsen Mostafavi and Gareth Doherty, Harvard University, Graduate School of Design, (Baden: Lars Müller Publishers, 2010).

25   See Marquis W. Childs, *Mighty Mississippi: Biography of a River*, (New Haven: Ticknor & Fields, 1982) 146–158 for an anatomy of a river town.

26   *Webster's*, 668.

27   Further information about the remix pile-up found on Lavie's bookcase:

   A.   Stan Allen and Marc McQuade, *Landform Building: Architecture's New Terrain* (Baden: Lars Muller; New Jersey: Princeton University School of Architecture, 2011).

   B.   Aude-Line Duliere and Clara Wong, *Monsterpieces,* (USA: ORO editions, 2010).

   C.   Jakob von Uexküll, "A Stroll Through the Worlds of Animals and Men: A Picture Book of Invisible Worlds," *Instinctive Behavior: The Development of a Modern Concept*, ed. and trans. Claire H. Schiller (New York: International Universities Press, Inc., 1957). See umwelt theory for understanding the life of animals through the "self-world."

   D.   Instructional DVD by Ferran Adria, the world's most awarded chef, and forms of food learning that have gone from "molecular gastronomy" to "avant garde cuisine" as touted in the Harvard course "Science and Cooking." See Ike DeLorenzo, "A First Course Gets High Grades," *The Boston Globe*, 29 December 2010.

28   See Charles Waldheim, *The Landscape Urbanism Reader* (New York: Princeton Architectural Press, 2006) and the Learning Channel's *Junkyard Mega Wars*.

29   Isabel Wilkerson, "The River Untamable," *The New York Times*, 08 May 2011: WK3.

30   *Webster's*, 629.

31   Suzanne Stephens and Clifford Pearson, "Millennium Part Two: Futures to Come" in *Architectural Record*, December 1999, 85–126. Interestingly, six of the nine projects proposed by firms for this journal were sited within or alongside bodies of water. See Asymptote, Museum of Technology (East River, New York); Michael Sorkin, Wheelchair Village (a generic Midwest river); Greg Lynn, Embryologic House (variable sites, earthen levee suggests the house could be placed alongside a body of water); Krueck & Sexton, Glass Tower (Chicago River); and Hariri & Hariri, The Cine-Pier #2 (East River, New York).

32   See http://www.fema.gov/plan/prevent/floodplain/nfipkeywords/base_flood.shtm (accessed February 2012). "The flood having a one percent chance of being equaled or exceeded in any given year. This is the regulatory standard also referred to as the "100-year flood." The base flood is the national standard used by the NFIP and all Federal agencies for the purposes of requiring the purchase of flood insurance and regulating new development."

33   On being energized rather than paralyzed by an agenda for renewal, especially during moments of economic melt-down, see William C. Taylor, *Practically Radical* (New York: HarperCollins, 2011): xi–xx. Taylor also makes the case for the human psychology part—or "animal spirits"—of the economy and illustrates how design in particular historically excels during turbulent times. Additionally, he references the Tom Friedman quote to follow, calling it the "mantra of the moment from the White House to Silicon Valley."

34   Thomas L. Friedman, "Kicking Over the Chessboard", *New York Times*, 18 April 2004. Economist Paul Romner coined the term "A crisis is a terrible thing to waste" referenced here by *New York Times* columnist Thomas Friedman.

10

Value Engineering and the Life Cycle of Architecture

George Barnett Johnston

One can look back a thousand years easier than forward fifty.

Edward Bellamy

*I believe that all education proceeds by the participation of
the individual in the social consciousness of the race.*

John Dewey

*In Mr. Palomar's life there was a period when his rule was this: first, to construct in
his mind a model, the most perfect, logical, geometrical model possible; second,
to see if the model was suited to the practical situations observed in experience;
third, to make the corrections necessary for model and reality to coincide.*

Italo Calvino

EXISTING CONDITIONS

In order to intervene in the world, to change it, to shape it for a better future
however one defines it, one must first be able to describe the world as it is. A
corollary might be that the unintended consequences of our interventions in the
world—and most human efforts amply result in these—are directly linked to the
relative crudeness of our models of the world, to their inadequacy in accounting
for unknown complexities and contingencies, in accounting for uncertainty. It
might even be expected that architects as professional agents so oriented by their
vocation to the reconciliation of existing conditions with future contingencies,
of change and use, would embrace uncertainty as a basic tenet of their practice;
that they would covet the potential of unpredictability as a creative catalyst in an
improvisational and reflexive process.

Yet it is only rarely so.

Rather than a steady driver with a ready arm on the reins, nudging unruly
forces by subtle moves along the unpredictable course of an unfolding path,
the architect is more often like a passenger on a careening stagecoach watching
disaster unfolding ahead. Reticent to risk liability for circumstances beyond their
own control, to be guarantors of price and performance, architects too often find

themselves as formal arbiters relegated to a subsidiary and reactionary role in the design process rather than one enterprising and forward leading.

Under prevailing professional conditions, it is more likely that architects, in grappling with complexity and concrete though shifting demands, are content to impose their wills upon the world in the form of partial and autonomously conceived models, ones to which the world (and users) must in turn accommodate themselves. Perhaps in the end "the world," our culture, is no more or less than the cumulative product and detritus of the unintended consequences that issue from—or are in spite of—all our best intentions. In the face of myriad human, social, political, economic, and environmental uncertainties, how could the formal certitude of any singular design agenda (whether it be the transformation of historical precedent or some generative geometrical logic) ever allay the profession's growing performance anxiety?

A dominant historical narrative holds that the profession of architecture in its modern form was forged in the Italian Renaissance; that the development of exacting representational standards within an intellectual climate of proto-scientific inquiry and intensified historical consciousness shaped the organization of artists' and craftsmen's workshops; and that those workshops in turn served the aggrandized visions of popes and princes, precursors of the corporate capitalists who emerged in the nineteenth century and came to dominate in the twentieth. The re-alignment of techniques and capital issuing from these developments both enabled and was enabled by a division of labor that ultimately yielded the basic structure of the modern profession that we recognize today. That profession can be characterized by its separation of design from production, by ingrained dialectics of architecture versus construction, and of theory versus practice. The result is a colonization of specialized spheres of expertise among architects and associated design disciplines as well as civil, structural, mechanical, and electrical engineers.

A prevailing consensus now holds that all of these dichotomies are under challenge by ongoing revolutions in digitally mediated design and fabrication; and that these transformations promise to be as epochal in our day for reshaping the architectural profession as were advances in representational technology and the reordering of design expertise in the Renaissance. Even more than developments in technology and political economy, however, I argue that the most profound transformation of the architectural profession in the modern era has been propelled by the separation of architectural education from architectural practice, by the division of architectural knowledge between the university and the architect's bureau.

From antiquity until the middle of the nineteenth century, the sites of architectural production were also the venues of professional reproduction and training. Just as architects' judgment was cultivated through precedent and vocational experience, so too were guiding principles of old reconcilable with the contingent lessons of everyday practice. As the authority of ancient treatises was progressively challenged by empirical evidence, however, the quest for scientifically-based architectural principles and universally valid design methods became the special reserve of academicized theory. The consequence of this innovation has been the institutionalization of a paradox in which the assumed role of theory is to fill up practice like an empty vessel rather than to recognize it as theory's endless font.

In contemporary institutional terms, the relationship between architectural education and architectural practice is generally understood as that of means and ends, in which case individuals' future professional paths are prepared by exposure to and exercise of the accrued body of disciplinary knowledge. A conservative interpretation of this equation would see the role of education as a conveyance of culture, of accumulated wisdom and experience, as preparation for the contingent realities of practice. Change is slow and cumulative, and the disciplinary traditions maintain some continuity even as they shift to assimilate new forms and ideas. A more strident view might hold that the role of education is to open students' eyes to the grip of a dominant reality and then to the necessity of subjecting it to critique; not to merely extend it, but perhaps to subvert it or overturn it or replace it with something new. Progressive change, from this perspective, emanates from the university, challenges the status quo of the profession, refreshes it and leads it in new directions.

The cogency of this model is suspect on at least two levels, however. On the one hand, university-based architectural education can only ever be a partial education in the field. In the sense that experience is both the source and validation for theory, the only true proving ground for architectural education is architectural practice, which is itself bound by assumptions in need of ongoing evaluation and critique. On the other hand, the manifold realities of architectural practice are in no compelling sense foreshadowed by the idealized, reductive models advanced by architectural education. Whether they be tradition-bound or future-focused, such models bear scant resemblance to the world of practice, subject as it is to the vicissitudes of so many complicating factors, so many competing expectations and shifting demands that no manner of academic simulation could adequately describe. And while the milieu of practice is shifting, so too is the context of the university itself being transformed by dynamic social, economic, and technological forces. Higher education is being challenged to account, in ways heretofore rarely imagined, for its relevance, its cost, its existence. The situation is not, in any manner of speaking, stable.

It is ironic, therefore, and perhaps a bit sobering to recognize that the emergence of the "culture of professionalism" and expertise that defines and enables our contemporary life is tied so explicitly as both cause and effect to the rise of the modern university. The articulation of a distinct body of knowledge separate from its application as vocation and the transmission of that knowledge through university-based education became distinct defining characteristics of the modern professions. Likewise, the incorporation of the field of architecture into U.S. universities beginning in the late-nineteenth century and accelerating after World War I reflected a host of related trends in the rise of business culture, especially the embrace of Fordist and Taylorist approaches to production including the division of labor and the specialization of knowledge.[1]

Doubts about the adequacy of this system issue to this day, however, from *both* the university *and* the profession. On the one hand, the university does not consider the fundamental knowledge and experience imparted in architecture school to be pure theory, in the sense of verifiable scientific principles and fundable research agendas. On the other hand, the profession does not consider architectural education to be particularly relevant to practice, as evidenced by complaints about recent graduates' inability to immediately contribute to firms' productivity and profit. The system of internship, influenced by both the overly romanticized atelier model of apprenticeship as well as the amply criticized system of articled pupilage, remains a necessary but somewhat awkward transitional period of supplementary training, of putting theory into practice. The intern is caught in-between, but is soon assimilated (when there is work available) to the practicing profession and so becomes a participant in reproducing the same maligned system. Idealists seek full-time faculty positions in the university and over time become more aligned with the university and its special demands than with the architecture profession itself. Thus the cycle is perpetuated.

Architecture in its broadest most potent sense is a metaphor joining vision and know-how. When perfectly aligned, the ends and means of architecture can serve grand purposes. Vitruvian principle assumed that the ennobling quality of beauty was the essence elevating mere building to an art. At the time of the American Revolution, architectures of antiquity were reinterpreted to bear the meanings of shared cultural values like personal liberty and social justice. Under current conditions, however, the means of architecture practice are undergoing such radical transformation that the intended ends of architectural education are increasingly obscure. Is architecture just another commodity undergoing a production line update, or can the new means of design simulation and component fabrication serve some more ambitious progressive cause? In short, has the full potential of architecture's role in advancing the health, safety, and welfare of the public been adequately explored?

ARCHITECTURAL VALUE

There is therefore, at the heart of things, an unresolved tension in U.S. architectural education and practice between old world and new world values, between patrician hierarchies and democratic ideals, between notions of tradition and progress. In the United States of the early-twenty-first century, issues of tradition and cultural value are often either deemed irrelevant to the dynamics of popular culture or else are subsumed by the rhetoric of conservative politics. These issues are, nonetheless, essential to the formulation of any critical practice of architecture. They penetrate to the core of architecture's relevance as a discipline while testing its continued economic viability as a profession.

Competition between motives of profit-making and benefits of cultural and environmental stewardship defines the socio-political context within which the value of architecture must be considered. Under the regime of cost benefit analysis, advocates of short-term profit and long-term investment each vie for the support of public policy and the conscience of private action, while in architecture, value engineering and life cycle costing have become euphemisms for this dialectic of economic determinism. And yet, it seems, it has always been so. Even Vitruvius acknowledges the existence in his time of the shoddy, the wasteful, and the ill-conceived, the work of charlatans and profit-mongers who give the profession a bad name.[2] Vitruvius is clear, however, about which characteristics he believes constitute the added-value of architecture and which comprise the ethos of the profession. *Firmitas, utilitas*, and *venustas* are not only standards of architectural quality; they are also public measures of private virtue—the patron's, the architect's, the builder's—in the stewardship of tradition and the renewal of cultural values.

We inherit from Vitruvius a model of architectural knowledge grounded in the liberal arts as we know them, seen from the standpoint of the 21st century and filtered through multiple epistemological lenses. What we make of Vitruvius, in other words, is fraught with interpretive jeopardy, vulnerable to misreading, and subject to debate. The sheer staying power, however, of his performance criteria of *firmitas, utilitas, venustas* merits our respect. Even from our privileged vantage point it is difficult to identify missing dimensions in that triad of architectural value. It is malleable in its applicability to different scales—from grand systems to simple structures. And while there is an evident objectifying bias of artifacts considered from their outward aspects and inner structuring logics, there is also the implication of their resulting effects, and their affects, and thus their entanglement with human subjectivity and collective purpose.

What firmness, commodity, and delight do not account for, however, is the whole process by which cities and buildings and artifacts are infused with those very performative attributes *De Architettura* promotes. Vitruvius' descriptive classification of architectural orders gives us scant insight into the process of design, or the implementation of designs—that is to say, the means by which imagination and invention are merged with labor and know-how to yield meritorious results. Vitruvius tells us what architecture should *be*, but he does not tell us what architects should *do*. The Vitruvian trinity gives us prescriptions based on precedents masked as principles. But what does it tell us about practice?

Vitruvius ascribes credit for the attainment of this trio of virtues, or the lack thereof, to the various parties involved with the work. If the project attains an overall magnificence, Vitruvius maintains, then the owner, by whose expenditures the work was accomplished, deserves recognition. If the workmanship is fine, then the builder must be lauded. "But," Vitruvius contends, "when [the building] has a graceful effect due to the symmetry of its proportions, the site is the glory of the architect. ...[For] the architect, when once he has formed his plan, has a definite idea how it will turn out in respect to grace, convenience, and propriety."[3] Though Vitruvius contends that public acclaim should be the architect's due reward for such attainments, he bemoans the fickleness of fate in so often bestowing public approbation based upon the architect's social influence. "Yet we must not be surprised," he says, "if excellence is in obscurity through the public ignorance of craftsmanship."[4]

Although Vitruvius is certain of the measures of architectural virtue, he seems resigned to their sporadic application and to the public's inability, due to lack of education, to distinguish the good from the bad. And he bemoans the ethical morass of his time, of architecture being practiced by those untrained and unqualified. Thus rationalizing for posterity his own lack of contemporary renown, he justifies his treatise-making and entrusts his principles to posterity. The positive model thus portrayed is of the architect as principled practitioner, a champion of architectural value, and a proponent of public virtue. But a shadow model is portrayed as well: of the architect as unscrupulous pretender, as social opportunist, and as a threat to public propriety.

In our own time, and continuing the tradition of Vitruvius, the various professional, pedagogical, and polemical purposes ascribed to the *Ten Books* have been assumed by a plethora of other handbooks, manuals, specifications, and treatises. Among these, the *Architects Handbook of Professional Practice* serves an indexical role in portraying the complexities of contemporary architectural practice within the techno-political context of the early-twentieth first century. Since its inception in 1917, the handbook has evolved from a quaint compendium of practice tips and standard forms to a comprehensive survey of professional topics, project procedures and tools, and contractual documents. Over the course of multiple editions, one can detect both a superficial adherence to the Vitruvian principles of architectural value as well as an evolving attitude toward the professional and ethical standards that society demands.

The preface to the early editions of *The Handbook of Architectural Practice* invokes the Vitruvian triad indirectly by reference to the duality of art and science within architecture. Regarding art, the authors write that "[i]t is as a fine art that architecture has established itself in the hearts of men. . . To good building, architecture adds qualities of the imagination. It disposes of masses and details in ways that arouse us by their beauty, power or dignity. It writes the record of civilization."[5] This view of art as the added-value of architecture is balanced by comments about the role of science within the architect's craft. "The Architect," they write, "though primarily an artist, must still be the master, either in himself or through others, of all the applied sciences necessary to sound and economic building, sciences that have generated and that attempt to satisfy many of the exacting and complex demands of modern life."[6]

In contradistinction to the purpose or effect of architecture, the role of the architect is made explicit in terms of professional virtue, for, according to *The Handbook*, "[t]he Architect, . . . by bearing himself as worthy of his high calling, gives to his art the status of a profession."[7] In another citation combining performative criteria for architecture with standards of professional performance, the sway of Vitruvian principle in *The Handbook* is made evident through an invocation against involvement in the erection of "unsafe, unsanitary, inconvenient, or unsightly structures."[8] Clearly, ancient and early-twentieth century opinions about professional responsibility coincide on the interest of the public trust.

To all of these characteristics of architectural value and professional virtue which the architect is bound to uphold, we must add the expectation that the cost of the building must fall within pre-ordained limits. Even Vitruvius rendered an opinion here, citing ancient Greek laws pertaining to the architect's personal liability for excessive cost over-runs.[9] *The Handbook* of 1927 states that "[o]ne of the Architect's most serious tasks lies in estimating the probable cost of the work."[10] Given the unpredictability of price due to the volatility of market forces, however, *The Handbook* seems less sanguine than Vitruvius in claiming this professional obligation and maintains that ". . . the Owner must in justice forbear hasty judgment if the Architect fail to display the gift of divination."[11]

The *Manual of Office Practice for the Architectural Worker* published in 1924 and adapted from the office manual of the firm of McKim, Mead, and White, links issues of construction cost to those of quality, describing a set of relative values for durability of construction, and even degree of ornamentation, based upon building use. Echoing Vitruvius' specifications for the requirements of defensive, religious, and utilitarian structures, a descending scale of construction quality and maintenance costs commensurate with durability is suggested for the buildings devoted to monumental, residential, and commercial functions. Of the latter, the manual flatly states that "[i]n all buildings of commercial character the element of income return is the most important factor. . .," while anticipating the logic of life cycle costing by asserting that the architect should have ". . . always in mind a minimum maintenance cost over a long period of years."[12]

So notwithstanding the pre-Depression era caveats, by the time of the 1943 edition of *The Handbook*, an AIA document promoting "The Value of an Architect" confidently states that "[a] good Architect often saves the Owner a sum much larger than his fee."[13] The 1958 revision goes even further in claiming "[the architect's] contribution to the work enhances the value many times more than the amount of his charges. Architectural service does not cost—it pays."[14] At the same time that claims for the profitability of architecture arise, commitment to the traditional Vitruvian triad appears to wane. The 1943 edition refers hesitantly to architecture's "claim to beauty" while emphasizing other issues of comfort, health, building knowledge, planning efficiency, and attractiveness. The 1958 edition dispenses altogether with references to beauty or even to attractiveness, substituting instead an uncertain notion of "distinctive design" while emphasizing "good building, economy, and efficient building operation." In an era of increasing programmatic and technological complexity, these AIA documents portray a profession within the throws of modernization, striving to promote its relevance and intrinsic economic value through problem-solving expertise and professional service. Concomitantly, re-orientation toward the precepts of cultural modernity precipitated the widespread adoption of such engineering criteria as efficiency and economy of means as substitutes for the classically precedented concepts of architectural beauty and civic propriety.

What *is* value? Lawrence D. Miles, the engineer at General Electric credited with the development of techniques of value analysis and value engineering, has succinctly and confidently addressed this bothersome question. Miles defines value as "the minimum dollars which must be expended in purchasing or manufacturing a product to create the appropriate use and esteem factors." He defines use value as "[t]he properties and qualities which accomplish a use, work, or service." Esteem value is defined as "[t]he properties, features, or attractiveness which cause us to want to own it." Value engineering, then, is concerned with the optimization of the use and esteem values of a product at the lowest possible price.[15] A gross analogy with the Vitruvian triad would suggest that firmness and commodity might be comfortably subsumed within Miles' notion of use value, thus assigning beauty to the bounds of esteem.

The design and building professions' adoption of value engineering concepts for purposes of cost control in construction is well-established today. Though the cost-benefits of such savings strategies for building are clear, efforts at engineering the esteem value of architecture are less easy to identify, much less to evaluate. Over the intervening decades since the Second World War, the debate about architectural value has, depending upon one's point of view, either inflated or collapsed. Certainly, concern for the initial and life cycle costs of buildings has intensified the quest for objective measures, in economic and environmental terms, of the relationship between durability and maintenance. The focus upon utility and cost, however, seems to be at the expense of clear postulates about the efficacy of beauty understood in any but the most subjective of terms, or else lodged within the instrumental domain of functional, legal, or economic parameters. In the face of such difficulty in defining the public virtues of beauty, commercial interests have subsumed art as the added-value of architecture. As esteem is conflated with novelty and the branding logic of a global marketplace, so too is architecture treated as just another commodity, as a vehicle for private profit and the pursuit of individual happiness.

REVOLUTIONARY POLITICS AND CONFLICTING CONCEPTS OF SOCIAL VIRTUE

Now, dragging Thomas Jefferson into this discussion may seem far-fetched, but the evidence suggests some relevant analogies. Volumes continue to be written about Jefferson's intellectual intentions with regard to the wording of the Declaration of Independence and its influence upon subsequent social, political, and economic developments within the United States. And given Jefferson's amateur status as an architect, like the ones Vitruvius praised in his own time as "those owners of estates who, fortified by confidence in their own erudition, build for themselves,"[16] it does not seem unreasonable to speculate about the coincidence of Vitruvian and Jeffersonian principles. In short, is it possible to see any parallel correspondence between, on the one hand, Vitruvius's *firmitas, utilitas,* and *venustas*, and on the other hand Jefferson's *life, liberty*, and the *pursuit of happiness*?

Conventional accounts of the philosophical genealogy of the Declaration of Independence locate the intellectual antecedents of Jefferson's construct of life, liberty, and the pursuit of happiness in the ideas of John Locke, most notably from the second of his *Two Treatises of Government*. These writings lay out a rationalization, through the appeal to natural law, of the basic human characteristics of freedom, equality, and independence. Within this conception of human nature, the individual is, according to one interpreter, "essentially a material being, ruled by the private senses which he shares with no one, guided by the pleasures and pains of this world, and motivated primarily by a desire for continued life, or 'self-preservation.'"[17] From this primal motivation for self-preservation, therefore, are derived the natural rights of humankind, namely life, liberty, and property.

In Locke's scheme, the emphasis upon the individual, and upon individual self-interest, outweighs the imperative for social conscience or virtue, which Locke considers to be an acquired rather than an inherent human trait. By nature independent, humans join together for pragmatic reasons. They form political organizations, for example, in order to preserve their individual freedoms against the threat of external forces. Thus, we understand the analogy that Jefferson constructs for the joining of the thirteen individual colonies to oppose the tyrannical impulses of George III.[18]

Much has been made of Jefferson's substitution of the phrase "pursuit of happiness" for Locke's notion of "property," even though Jefferson's ideas about property as the vehicle for economic independence seem to conform to Locke's own attitudes toward the psychological motivation of happiness. Locke suggests that at the root of so many diverse and deeply-held moral convictions among differing peoples is a common human aspiration toward happiness. This concept of happiness is not limited to the immediate pleasures that obscure through sensation the memory of past pain. Rather, he maintains, true happiness derives from the consciousness that one has amassed the means in the present, in the form of property and the power over it, to render pleasure in the future and thus to avert future pain.[19] The social implications of these concepts are devastating, for they assume that humans are naturally solitary, that they join together only under common threat for purposes of self-preservation, and that, beyond safeguarding the individual from the infringements of others' actions, objective standards for the conduct of human affairs cannot transcend the realm of subjective choice.[20]

During the late-eighteenth and early-nineteenth centuries, extension of these same enlightenment concepts into the philosophical domain of architecture resulted in strikingly similar conclusions. Especially in the theoretical writings of J.-N.-L. Durand, a pedagogue of the French *École Polytechnique*, Alberto Perez-Gomez has observed the emergence of a new value system founded upon complementary principles: "love of well-being and aversion to pain." Perez-Gomez notes that "[t]his materialistic premise became the basis of the ethics and aesthetics of technology, and it still underlies the most popular historical and ideological conceptions inherited from the nineteenth century. Only after Durand would it become important for architecture to provide 'pleasure' or that it be 'nice' rather than truly meaningful."[21]

Durand's transformation of Vitruvius is equally significant. In place of the traditional triad, Durand substitutes an emphasis upon economy and efficiency, stressing the conformance of architectural principle to the reason inherent in natural law. According to Perez-Gomez, "[t]he system of values in architecture was thus reduced to a scale between pleasure and pain. Value could be 'measured' as it approached ideal efficiency and maximum pleasure." As for beauty, Durand rejected all ornamental embellishments to architecture, save those derived as a result of the dictates of convenient and economical disposition of the plan. Again according to Perez-Gomez, "[t]his system of values lays at the origin of the still prevalent emphasis on comfort over meaning in contemporary architecture."[22] Such explicit foreshadowing of the precepts of value engineering, of the optimization of "use value" along with the linking of "esteem value" to the whim of personal desires, confirms, I believe, certain parallels between enlightenment notions of liberty and our own departures from Vitruvian principles of architectural virtue. Is this really Thomas Jefferson's contribution to the formulation of an American architecture?

Not necessarily. Revisionist historiography since the 1960's has challenged the unity of Lockean interpretations of Jeffersonian intent in crucial matters surrounding the Declaration of Independence. While acknowledging Jefferson's deep debt, verging on plagiarism, to Locke's *Second Treatise,* those critics of the Lockean view point toward Jefferson's equal erudition on the concepts of classical virtue.[23] Besides the connoisseurship of classical architecture predating his European travels, Jefferson was obviously well read in classical philosophy and literature. A difficulty thus arises when one tries to reconcile the pure revolutionary zeal for the rights of the individual expressed in the *Declaration of Independence* with his post-revolutionary commitment to the nurturing of republican social values within the fledgling democracy. In contrast to the imperatives of individual liberty emphasized by Locke, classical social ethics, especially as expressed by Aristotle, insists upon a standard beyond merely not harming one's neighbor; rather, it was incumbent upon the individual to assume the moral responsibility for the improvement of the plight of others within the framework of objective standards either revealed by God or else developed through collective agreement.[24] For Locke, human beings were naturally solitary and independent; for Aristotle, man was a social and political animal.

Evidence for Jefferson's adherence to a concept of happiness contrary to the pecuniary interests of Lockean property are found in his own writings. For example, in a 1788 letter, Jefferson advises his nephew that "[h]ealth, learning, and virtue will insure your happiness; they will give you a quiet conscience, private esteem and public honor." This advice would seem to echo Aristotle's equation of the purpose of politics, that is "the highest good attainable by action," with the concept of happiness.[25] Jefferson's notion of happiness and its pursuit transcends the private pleasure of Lockean property to include the public acclaim of good works. So too, as accords to Vitruvius, who wrote that ". . . all men, and not only architects, can approve what is good."[26]

Whether Thomas Jefferson owned a copy of Vitruvius prior to the commencement of the work at Monticello is unclear, but it is known that he did possess an edition of Palladio's treatise which would have nurtured his familiarity with the more ancient text.[27] The preface to Palladio's treatise articulates a clearly classical concept of social virtue, describing ". . . a man, who ought not to be born for himself only, but also for the utility of others."[28] And in his *First Book*, Palladio refers immediately to the Vitruvian triad of architectural virtues, commenting "[t]hat work therefore cannot be called perfect, which should be useful and not durable, or durable and not useful, or having both these should be without beauty."[29] Is it too outlandish to suggest that for Jefferson classical notions of architecture were metonymically linked with enlightenment concepts of human nature: that preservation of life was dependent upon a building's durability, that liberty was sympathetic with principles of utility and propriety, or that the pursuit of happiness could be lodged in the aspiration toward beauty and its public acclaim?

The oscillation and cross-pollination between republican and liberal political ideals, in this way, is exactly what some authors claim Jefferson contributed to the formation of the new constitutional system: revolutionary and libertarian against the concentration of central power, classically republican in the advocacy of local participatory democracies.[30] How this *rapprochement* between individual freedom and social conscience has evolved over the course of two centuries can largely be described in terms of intertwined technological, economic, and popular cultural developments, each of which has exerted its transformative power upon the discipline of architecture as well. While in society at large, technology has enhanced individual freedom and well-being through strides in fields including transportation, communication, construction, and medical science, it has also contributed to the degradation of the environment, the fragmentation of experience, and the instrumentalization of value. Economic self-interest has bred opportunity and independence as well as discrimination and dependency. And the influence of popular culture stretches the bounds of human imagination even while engendering a profound cultural homogeneity.

Architecture has been swept by these tides as well. Jefferson's ideals of governance—life, liberty, and the pursuit of happiness—over time and with the application of the Tenth Amendment to U.S. Constitution, have translated into the public virtues and personal protections of *health, safety, and welfare*.[31] The establishment of the state-regulated profession of architecture in the United States has begotten the legal safeguards, regulatory bodies, and collateral organizations which constitute the framework of contemporary architectural practice. Thus, the trajectory which has led us from the Vitruvian tradition of *fimitas, utilitas, venustas* to the contemporary legal standard of health, safety, and welfare is fraught with contradictions. By accepting technological, functional, and economic criteria as defining the limits of its legal, and thus civic, responsibility, the architectural profession has unwittingly promoted private, commercial interest as the one standard for its public validation. The public virtue ascribed to Vitruvian beauty, which remained at least credible as a private motive for the Jeffersonian pursuit of happiness, has assumed a decidedly Lockean character in the linkage of public welfare with the pursuit of private wealth. The *art* of architecture, that transcendence of mere utility which defined architectural value from ancient times, is thus reduced to the esteem of fashion or the rhetoric of formal and theoretical preoccupations.

But if Vitruvius's simple algorithm of value has not survived unscathed, neither have the values engineered by Jefferson. Jefferson's post-revolutionary commitment to local republics and decentralized economies has simply not prevailed. His principled appeal to the pursuit of happiness through public virtue has instead been taken as a justification for self-indulgence. This profound misreading of Jefferson lies, I believe, at the heart of many national dilemmas, and it contributes to the crisis of value which the profession of architecture endures. While health, safety, and welfare suggests the broad bounds of a public trust, it is, in the end, only defined by so many private interests.

THE SOCIAL VALUE OF ARCHITECTURE

As two sides of the disciplinary coin, architectural education and architectural practice are joined in an uncomfortable co-dependency. They both grapple with foundation-shaking challenges in technology, society, and the environment while struggling to imagine a way forward, a path beyond the trite caricatures that each has painted of the other. Education is key in both Vitruvian and Jeffersonian schemes for ensuring the union of public and private virtue. In the early-twentieth century, John Dewey strongly advanced the notion that "education is the fundamental method of social progress and reform."[32] Professional education is key, then, not only to the qualification of architects as practitioners of the status quo but also in preparing them to serve as midwives of change. Following in the spirit of these, we should add that the culture of the built environment ought to be a significant focus of general education—in order to inculcate public appreciation of architecture, of the role it plays as a vessel of social progress and cultural value, and of the role individuals themselves can play as agents shaping public and private domains.

So if there is an essential connection between education and social reform, then what is the social agenda of architectural education?

Besides offering metrics for assessing architectural performance, Vitruvius inventories a long list of subjects that he argues should constitute the architect's education, and presumably the public's, and thus provide the underlying knowledge enabling the realization (and appreciation) of firmness, commodity, and delight. "Let him be educated, skillful with the pencil, instructed in geometry, know much history, have followed the philosophers with attention, understand music, have some knowledge of medicine, know the opinions of the jurists, and be acquainted with astronomy and the theory of the heavens." Anticipating readers' incredulity "that human nature can comprehend such a great number of studies and keep them in memory," Vitruvius invokes a corporeal metaphor of knowledge arguing, "For a liberal education forms, as it were, a single body made up of these members"; and "all studies have a common bond of union and intercourse with one another."[33] Architecture is portrayed in the same analogous terms, shaped in part and whole by reference to a paradigm of bodily proportion.

Among the nine essential subjects described by Vitruvius, it is possible to see both reverberations of Aristotelian philosophy and a foreshadowing of Medieval Scholasticism. Of the seven subjects, I would argue that eight are devoted almost exclusively to the challenge of describing the world, thereby to provide a basis for acting within it. They establish the necessary tools and representational systems—

letters, numbers, lines—for surveying existing socio-political conditions and for correlating them with physical, environmental ones.

Letters. Being himself a treatise writer, Vitruvius prioritizes mastery of letters as an essential trait, thus defining the architect as a learned individual capable of both being informed by and informing others about all that has been accomplished in the field. To be educated in letters, however, implies a much broader purchase than simply the ability to read and write with fluency. What is more fully suggested here, as considered, say, within the scope of the medieval trivium—of grammar, logic, and rhetoric—is a capacity for the formulation of thought and exercise of reason through language. That is to say: the ability to make persuasive arguments, and to follow and critically assess those made by others. And if language is the structuring medium of thought, it is also a mechanism and model for culture: the interplay of syntax and semantics, the signification of meaning and its shifting figurations, the simultaneous precision of objects within an unstable flux of possible interpretations.

Vitruvius further specifies knowledge of history, philosophy (both moral and natural), and law as being important components of the architect's education for which the study of letters is of course essential preparation. Assimilation and reflection upon the lessons offered by each accounting of the world requires fluency in language and close study and comparison of cases; but exposure commences, in a practical sense, from childhood. It comes through so many stories featuring the exploits of heroes; moral parables about choices and their broader consequences; the direct experience of cause and effect in nature, of authority asserted and then rebuffed, of public and private interests. What is at stake in this sort of liberal education is the formation of the neophyte's judgment in preparation for practice, in advance of any substantial professional experience. Liberal education of this sort distills the accrued experience of many cultural practices, essentializes them and gives them back to serve as an inoculation for life.

Numbers. In Vitruvius, quantitative and qualitative value are co-extensive rather than being diametrically opposed. From arithmetic calculation of area, volume, material, labor, and cost; to the analogical equation of bodily and building proportion; mathematics permeates architecture through and through. Beyond the pure and associational values of numbers, number is spatialized through geometry, the system for describing one, two, and three dimensional entities—points, lines, surfaces. Besides the visual proofs of abstract spatial relations, geometry provides explanatory power for optical principles of visuality and the properties of light; and by providing the underpinning for drawing, geometry instructs the union of the visible and invisible realms in and through representation.

Vitruvius' inclusion of astronomy and music with arithmetic and geometry as essential features of the body of architectural knowledge completes this curricular quadrivium. Music embodies numerically ordered tonal relations unfolded over time within geometrically reverberant acoustical space. And astronomy provides a model for grasping the interplay of celestial and other bodies in space over time, dynamically interacting from multiple relational perspectives. Numbers provide a multivalent means for describing the world in objective and comparative terms, ones supplementary to letters but which when combined with them enable us to grasp relationships, and the relationships among relationships.[34]

Lines. Vitruvius' simple injunction that the architect should be "skillful with the pencil" belies the profound role that visual representation plays in the constitution of the discipline of architecture. In the same way that we accept that thought is shaped within the medium of language, or that mathematics is the language of nature, so too do visual media provide a cognitive bridge between eyesight and insight, between memory and imagination, vision and know-how. The three species of architectural drawings that Vitruvius describes—*ichnographia, orthographia, scaenographia*—can be employed as a means of mapping existing conditions of the physical realm, proceeding from the world in its fullness through objectifying filters that both clarify and collapse experiential differences into geometric abstractions. The act of drawing the world, of measuring it, already alters it, reconstructs and prepares it as a site for subsequent interventions. Architectural design, then, implies an inverse operation in which lines are projected into surfaces and then surfaces into the constructed forms and spaces by them defined.

Architects' working drawings have typically combined all three descriptive orders of lines, numbers, and letters into a composite representation, dimensioned and notated drawings supplemented by written contracts and specifications. The rise of digital building information modeling further enhances our capacity for describing complexity, for simulating performance. Its greater emphasis upon geometric and mathematical representation suggests a concomitant decline, however, in the explanatory force of letters (and the liberal arts) in the face of computer numerical control. Under prevailing technologies, reading and writing as continuous practices have been undone by hypertextuality. The "jump" of the hyperlink to non-validated "external sources" is a far cry from the encyclopedists' ethos of learning-in-a-circle through authoritative cross-reference. That closed-loop system was designed to expand awareness, whereas hypertext promotes stream of consciousness in an attention deficit society. Still, such retrograde assessments might simply exemplify nostalgic, vestigial values waning on the cusp of some emergent manifestation: knowledge not as a body, whole and proportionate, but knowledge as a network, pervasive, dispersed, emergent, infinitely accessible, manipulable, and constructed.

Practices. The final domain that Vitruvius invokes as part of the architect's requisite knowledge is medicine. "The architect should also have a knowledge of the study of medicine on account of the questions of climates (in Greek *klimata*), air, the healthiness and unhealthiness of sites, and the use of different waters. For without these considerations, the healthiness of a dwelling cannot be assured." By these examples is suggested a range of methods by which diagnostic inferences derived from examinations of the world can be translated into compensatory design prescriptions. Vitruvius' citation of medicine conjures a whole realm of action analogous to the practice of architecture within which principle, evidence, and circumstance coincide in the field of creative architectural production, and then in the assessment of its efficacy. The same sorts of diagnostic tools and imaging devices that have revolutionized medical practice now permeate architects' practices as well. In neither case, however, can those tools substitute for judgment. While letters, numbers, and lines are basic components of architectural literacy enabling the reading and writing of architecture culture, and while each in turn may serve as the frame and focus of architectural research and speculation, in the end architecture knowledge, like medicine, is only constructed case-by-case through situated action. Only inasmuch as we truly inhabit the world are we able to practice architecture responsibly.

Architecture is shaped by the body, and by reference to it; it is infused by the body and is a manifestation of it, a constructed body of knowledge, now extended and dispersed into a vast network of potential neural connections. In this light, it might be worth considering whether the safeguard and enhancement of cultures', of societies', of individuals' health, safety, and well-being might not be, after all, the raision d'être of the profession, one tantamount to the medical profession's own Hippocratic Oath. Broadly interpreted, is it not the architect's responsibility to provide through spatial and environmental means across a range of scales the physical and psychological constituents of human health and happiness and to prioritize their pursuit? The embrace of this simple precept would advance by strong measure a renewal of architectural value and reorient architectural education toward a fundamental goal: providing students and future architects with the knowledge, tools, and sensibilities that they need to shape ennobling spaces for everyday practices and enabling settings for social reform.

Architectural practice and architectural education must be reintegrated in substantial ways to advance this goal. Old dividing lines must be blurred so that the education of architects is again understood as the responsibility of the profession, not the specialized domain of professors. What the university can do best is frame the intellectual context of liberal education in and through the disciplinary lens of architecture and then the methodological rigors of advanced research into the import of the constructed realm. Thus bracketed and bolstered, professional education must be firmly situated in the world, in the messy reality that the profession itself faces every day. Professional education should again be subsumed within the profession it reproduces. Practice, in its turn, must internalize the university's culture of questioning and critique, of research and discovery in order to constantly renew itself based upon precise questions and clear propositions about evidence and causality, meaning and significance.

I believe we must do all this in the interest of perfecting the profession and promoting its ability to serve the larger good. That of course includes enhancing professional productivity and performance. But that is not all. We must do this in order to cultivate a broader public understanding about what is at stake in the shaping of the built environment and then to inspire individual participation in a process of progressive change. Architects can only serve society if they are as attentive to enduring needs as they are open to new possibilities. Likewise, desire for social progress ought to be at the heart of architects' aspirations to derive truly progressive architectures, ones equally transformative of practice, use, and form. In contentious times, however, the measures of progress can seem difficult to define and impossible to agree upon. The very concept of progress is itself open to debate, ambiguous, perhaps even anachronistic in this age of perpetual novelty. Things don't get better, some would argue; they just get other. In uncertain times like these, I maintain, the difficult question of how to gauge social progress—in and through architecture—must be a central focus of our inquiry, the main objective of our creative and productive work. Architecture is a form of social vision embedded in things made as well as in the spaces we shape and the cultural practices they enable, and from which a vital, healthy culture can only emerge.

NOTES

1   See, for example, Burton J. Bledstein. *The Culture of Professionalism: The Middle Class and the Development of Higher Education in America*, (New York: Norton, 1976).

2   Marcus Vitruvius Pollio, *De Architectura*, trans. Frank Granger (Cambridge, Massachusetts: Harvard University Press, 1983), Book VI. Preface. 5–6. I keep open on my table two other translations of Vitruvius for constant comparison: that by Morris Hicky Morgan (Dover, 1960) and the more recent and pedagogically-focused illustrated version by Ingrid D. Rowland and Thomas Noble Howe (Cambridge University Press, 1999). I have not yet tackled, however, Richard V Schofield's translation (Penguin, 2009), but it is on my list.

3   Ibid., Book VI. Ch. VIII. 9–10.

4   Ibid., Book III. Preface. 3.

5   American Institute of Architects, *The Handbook of Architectural Practice* (Washington, D.C.: AIA, 1927), 11.

6   Ibid.

7   Ibid.

8   Ibid., 17.

9   Vitruvius, Book X. Preface. 1–2.

10   *Handbook* (1927), 44.

11   Ibid., 45.

12   Frederick J. Adams, *Manual of Office Practice for the Architectural Worker*, (New York: Charles Scribner's Sons, 1924), 52–53.

13   American Institute of Architects, *The Handbook of Architectural Practice,* (Washington, D.C.: AIA, 1943), Appendix S.

14   American Institute of Architects, *The Handbook of Architectural Practice,* (Washington, D.C.: AIA, 1958), A 7.03.

15   Lawrence D. Miles, *Techniques of Value Analysis and Engineering*, (New York: McGraw-Hill Book Company Inc., 1961), 3.

16   Vitruvius, Book VI. Preface. 6.

17   Garrett Ward Sheldon, *The Political Philosophy of Thomas Jefferson*, (Baltimore: Johns Hopkins University Press, 1991), 9.

18   Ibid., 45–52.

19 Thomas L. Pangle, "The Philosophic Understandings of Human Nature Informing the Constitution," in *Confronting the Constitution: The Challenge to Locke, Montesquieu, Jefferson, and the Federalists from Utilitarianism, Historicism, Marxism, Freudianism, Pragmatism, Existentialism...*, ed. Allan Bloom (Washington, D.C.: The AEI Press, 1990), 49–58.

20 Sheldon, 13–14.

21 Alberto Perez-Gomez, *Architecture and the Crisis of Modern Science* (Cambridge, Massachusetts: The MIT Press, 1983), 299.

22 Ibid., 303.

23 Sheldon, 14–15.

24 Ibid.

25 Ibid., 56.

26 Vitruvius, Book VI. Ch. VIII. 10.

27 Frederick Doveton Nichols, "Jefferson: The Making of an Architect," in *Jefferson and the Arts: An Extended View*, ed. William Howard Adams (Washington: National Gallery of Art, 1976), 164.

28 Andrea Palladio, *The Four Books of Architecture* (New York: Dover Publications, Inc., 1965), Author's Preface.

29 Ibid., Book I. Ch. I.

30 Sheldon, 17.

31 The Tenth Amendment establishes the principle of federalism reserving to the State the authority to extend its police power for the protection of the health, safety, and welfare of the public through the creation of laws. It reads: "The powers not delegated to the United States by the Constitution, nor prohibited by it to the states, are reserved to the states respectively, or to the people."

32 John Dewey, "My Pedagogic Creed," in *School Journal* vol. 54 (January 1897), pp. 77–80.

33 Vitruvius, Book I. Ch. I.

34 Kline, Morris. "The Sine of G Major." in *Mathematics in Western Culture*. (Oxford: Oxford University Press, 1953), 287.

11

The Architecture of Innovation

Thomas R. Fisher

This we know for certain: the need for shelter will continue to grow with the human population, which has increased at an exponential rate since the 19ᵗʰ century. It took us all of human history to reach 1 billion people globally in 1830, a number that doubled in just 100 years to 2 billion in 1930. In another 30 years, in 1960, we reached 3 billion people; 15 years later, in 1975, we reached 4 billion; 12 years later, in 1987, 5 billion; 13 years later, in 2000, 6 billion; and 11 years later, in 2011, 7 billion. By 2050, experts expect us to top 9 billion people, a four-and-a-half fold increase in just 120 years.[1]

While those numbers reflect better healthcare and increased longevity in some parts of the world, they also arise from the grinding poverty in other areas, where having a lot of children serves as a hedge against destitution. This exponential growth in our numbers also raises questions of how long the planet can sustain such a large human population, as we have already exceeded the carrying capacity of the globe, according to some assessments. And therein lies the great question of our age. Will we be able to raise the standard of living of such a large human population without crashing the resource base and altering the climatic conditions upon which human civilization depends? (Figure 11.1)

Architecture has an absolutely central role to play in answering that question, and so far, the architectural profession has done relatively little to address it. For sure, sustainability has become a goal of many architects, well aware that, as *Architecture 2030* argues, buildings consume almost half of the energy we use and the carbon dioxide we emit and roughly three-quarters of the electricity we consume.[2] At the same time, the ways in which we have developed the land have so altered and fragmented the habitat of other species that we have spawned what biologists call the "sixth extinction," in which about half of all of the species currently on the planet will disappear this century.[3]

A much smaller group of architects have also embraced the needs of the world's rapidly growing population. Landmark exhibitions such as the Cooper Hewitt's *Design For (and With) the Other 90%* and non-profit firms such as Architecture for Humanity and MASS Design and educational efforts such as Design Corps and Project H are a few of many efforts in the "public-interest" or "social-impact" design

11.1   Our Ponzi Scheme with the planet has led to exponential growth in our consumption of resources, now extending beyond the capacity of the planet to sustain. Ponzi Schemes collapse when they outrun their base of support and we are at that point. Kamana Dhakhwa

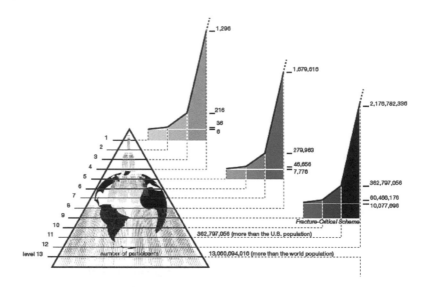

sector.[4] The fact that all of these efforts represent small-scale initiatives or one-off projects no where near enough to meet the needs of a burgeoning population remains a challenge as of yet unaddressed.

No architect need worry about a lack of work to do in the world. With the current human population already living beyond the carrying capacity of the planet, and with North Americans requiring almost 5 planets to meet our average consumption levels, we have long passed the point of tweaking existing systems in order live within the ecological footprint of one planet.[5] And with about 100 million people around the world homeless and an estimated 889 million who will be living in slum conditions by 2020, the need far outstrips the current level of pro bono efforts by the global architectural profession.[6] We face an immediate future in which almost everything we have inherited from our resource-intensive and economically exploitative past needs rethinking and redesigning.

The architectural profession, itself, needs the same. Most architects follow a "medical" model of practice, generating custom responses to individual needs of clients able and willing to pay professional-level fees. That has worked well for the world's wealthy who can afford such rates. However, the profession's focus on roughly the richest 10 percent of the world's population has led to its ignoring the needs of the other 90 percent who arguably need the services of architects even more. The architectural profession's attention to the wants of the wealthy has also encouraged the over-consumption of limited resources such as fossil fuels, clean water, and non-renewable assets that future generations and other species will need. Just because the rich can afford to do what they want doesn't mean the planet can afford it.

The great opportunity that the architectural profession has yet to grasp is the rise of a "public health" version of practice. Just as medicine gave birth to public health to address the health needs of the global population, the time has come for the architecture and design community to develop an equivalent path to meet the shelter, sanitation, and settlement needs of all the people on the planet. Public health has thrived since its emergence from medicine in the mid-19ᵗʰ century. And a public health version of the design disciplines will do the same given the billions of dollars that flow through foreign aid and development channels and the billions of people in need of low-cost, high-impact, resource-conserving solutions to their daily requirements.

As with public health, the clients of these public-health design fields will be less often individuals and corporations and more frequently entire countries, large non-profit organizations, major government agencies, and non-governmental entities of various sorts. And a public-health practice will involve not custom response to individual needs, but instead, prototypical designs for large groups of people appropriate to their culture and their social, environmental, and economic capacities.

The redesign of the architectural profession in terms of public health does not arise out of some bleeding-heart liberalism. As we have seen with the medical version of public health, we all benefit—rich and poor—from clean water, safe food, and clear air, and no amount of wealth will insulate us against the effects of an unhealthy environment. The same reality pertains to the architectural and design aspects of life. Too many of us, especially in the wealthier parts of the world, believe that we can protect ourselves from the catastrophic events that have begun to occur with ever greater frequency as our population numbers and our planetary impacts have increased. We think that we can insulate ourselves from the worst effects of our changing climate or from the instability of impoverished populations in other parts of the globe, but that is an illusion.

Deadly viral infections for which no cure exists typically arise in impoverished communities with poor sanitation that can lead to the transfer of disease from animals to humans or to the mutation of less deadly disease to more virulent varieties. And as the recent H1N1 epidemic showed, those infections move first to large cities with active international airports and to wealthier populations that travel more frequently and so come in contact with a greater diversity of people.[7] While H1N1 proved to be very mild, we may not be so lucky next time. The best way to prevent a global pandemic that could wipe out most humanity is to prevent such an outbreak from occurring in the first place through dramatic improvements to the living conditions of the world's most impoverished people.

Nor will anyone, however wealthy, avoid the devastating effects of climate change. While many people may welcome warmer winter temperatures, the real effects of a rapidly changing climate will manifest themselves in myriad ways, from increasingly violent storms to witheringly long droughts to devastatingly large floods. Indeed, some of greatest impacts of climate change will happen first in places where the wealthy often like to live or vacation—along low-lying coastlines, in warm "sun belt" communities, and in tropical resort areas.[8] Ultimately, we all live on the same planet and without several other "Earths" out in space ready for us to retreat to, we have only one option: learning to fit on this one world.

WHAT WE NEED TO DO

In learning how to fit on this planet, though, we encounter a paradox. Despite the enormous amounts of information now available to us in the Internet age, we have forgotten much of what our ancestors knew. Never before have we had so much data and so little understanding. A key aspect of human life we have forgotten involves how to live a happy life with very little in terms of material possessions. Most of humanity, of course, once did so and yet most of us, especially in the wealthiest countries, have no idea how to sustain ourselves on very little or live off the land.

We have become not only the most powerful of all the species on the planet, but also the most vulnerable, more dependent than ever before on processes and systems that are, themselves, liable to unpredictable interruptions or catastrophic crashes if a key resource become unavailable or an essential element fails in some way. We have designed a "fracture-critical" world for ourselves, subject to sudden and often disastrous collapses, as we have seen recently in everything from levee failures in New Orleans to the nuclear reactor meltdown in Fukushima to the investment banking collapse on Wall Street to the housing bubble burst in the U.S. to the debt-ridden implosion of the Greek economy.[9] The list could go on, and the story is the same: we have created incentives that encourage people to push systems to the point of failure, at which point, we all suffer. (Figure 11.2)

For no other reason than that, we need to redesign the environments and systems upon which we depend in order to protect ourselves from such possibilities. The few who benefit from this "disaster capitalism," as Naomi Klein has aptly called it, will do all they can to protect the system as it is.[10] They have shown themselves

11.2 Fracture-critical systems, like the original I-35 W bridge, have so little resiliency that they collapse suddenly once the stress on them pass a tipping point. This has become a pervasive problem and shows why we need to build in more resiliency into systems

adept at equating their exploitative practices with personal liberty and at winning over the very people often victimized by their tactics with the idea that everyone should have the same opportunity to exploit. We can only guess what Adam Smith, the father of capitalism and a professor of moral philosophy, would have thought of such outrageously immoral practices in name of a supposedly free market.

What, then, might a non-exploitative, non-fracture-critical future look like, one that enables us to live within the carrying capacity of the planet without undermining the very things we need in order to survive as a species? That may seem like a loaded question, but in answering it, we will rediscover a lot of what we once knew. And in so doing, the architecture and design community may come to see the new role it has to play in shaping a world that we can sustain and in which we can thrive for generations to come.

So, instead of trying to modify the world as we have designed it, let's start by engaging in a thought experiment in terms of what we humans might learn from the rest of the natural world of which we are a part. To do this, consider the discovery of the bio-physicist Geoffrey West and a group of his colleagues at the Santa Fe Institute. They have shown how most living beings exist according to constant relationship between mass and metabolism, matter and energy. To understand what that means, consider the following two equations. The one, based on physics, has come to represent the 20th century: Albert Einstein's $E=MC^2$. Encapsulating the relationship of matter and energy, that equation also epitomized the last century's pursuit of power, speed, and acceleration, which has helped fuel some of the exponential increases we have seen over the last 100 years. The other equation, based on biology, may come to represent the 21st Century: the Santa Fe Institute's $E=M^{3/4}$. As Geoffrey West describes it, "the metabolic rate varies as mass raised to the ¾…man is a little less than 100 watts in metabolism (a light bulb) – that's about 2000 calories a day."[11]

That relationship of metabolic energy to the mass of a living entity times the ¾ power apparently applies to all animal and plant species, with the exception of humans, whose use of technology has massively increased our absorption of energy. Each of us now has the metabolism equivalent, according to West, to that of a blue whale.

> How much energy does our lifestyle [in America] require? Well, when you add up all our calories and then you add up the energy needed to run the computer and the air-conditioner, you get an incredibly large number, somewhere around 11,000 watts. Now you can ask yourself: What kind of animal requires 11,000 watts to live? And what you find is that we have created a lifestyle where we need more watts than a blue whale. We require more energy than the biggest animal that has ever existed. That is why our lifestyle is unsustainable. We can't have seven billion blue whales on this planet. It's not even clear that we can afford to have 300 million blue whales.

The great challenge of our time entails reducing the average human metabolism from 11,000 watts (220,000 calories) per person to 100 watts or 2000 calories. That may seem impossible, but only if we assume that the last hundred years—

1/1,000th of our existence as a species—is the norm and all the rest of our history an aberration. Once we realize that humans long thrived on 100 watts a day, we can begin to imagine such a world again in the future, one in which people used tools rather than machines, creative power rather than nuclear power, renewable rather than non-renewable resources, and food rather than oil as a primary fuel.

This does not mean, though, that we need to return to what may sound like a primitive existence. Quite the contrary. West and his colleagues distinguish between the "sublinear scaling" of nature, in which the metabolism or pace of biological activity decreases with an increase in the mass or size of an organism and ecosystem, and "superlinear scaling" of human communities, in which the metabolism or pace increases with size, as in a city. As West writes,

> The bigger the organization, the faster the pace of life. In big cities, disease spreads more quickly, business is transacted more rapidly, and people walk faster —all in approximately the same systematic, predictable way… To sustain such growth in the light of resource limitation requires continuous cycles of paradigm-shifting innovations …There is, however, a serious catch: Theory dictates that the time between successive innovations must get shorter and shorter … Until recent times the interval between major innovations far exceeded the productive life span of a human being. But this is no longer true: The time between the most recent major shift from computers to IT was only about 20 years and is destined to get even shorter. This pace is surely not sustainable and, if nothing changes, we are heading for a major crash—a potential collapse of the entire socioeconomic fabric. Can we return to an analogue of the sublinear, "biological" phase whence we evolved and its attendant, natural, no-growth, asymptotically stable configuration?[12] (Figure 11.3)

11.3   The work of Geoffrey West and his colleagues at the Santa Fe Institute have shown that, as population increases, so does our use of resources (1) and if we are to avoid a collapse (2), we need to increase the amount of innovation (4) and the pace of it (7) in order to sustain ourselves on the planet

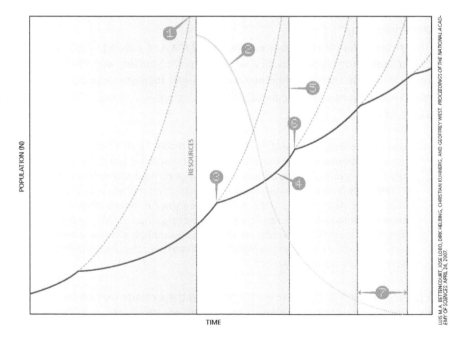

West's analysis of the "superlinear scaling" of cities raises the question: can we innovate at an ever-faster rate in order to avoid a "collapse of the entire socioeconomic fabric?" Can we, in restating our initial question, raise the standard of living of seven and soon to be, nine billion people through creative re-invention without crashing the resource base and altering the climatic conditions upon which we depend? Is there a limit to how fast we can innovate or at least absorb the innovations that arise or is human creativity and imagination a truly infinite—and largely untapped—resource?

We frequently view progress in terms of modern technology, but real progress in the future may involve, instead, the rediscovery of what humans once knew, but that we have forgotten or chosen to ignore. The great collective task we face involves finding the modern "analogue of the sublinear, 'biological' phase whence we evolved," as West put it through the "superlinear" uniquely human activity of creativity, discovery, and imagination. This involves not returning to a primitive past, but instead, relearning how to live within our biological limits as the next phase in the evolution of our species. This will take extraordinary inventiveness in order to achieve, and not just in the area of science and technology, but also in the structures and systems we construct for ourselves.

WHERE WE GO FROM HERE

This has already begun to happen in some respects. The Organization of Economic Co-operation and Development (OECD) has shown how those countries that have the least resources or depend least upon resource extraction have the highest student education scores, suggesting, as Thomas Friedman put it, that we need to "pass the books, (and) hold the oil."[13] In other words, when we stop depending on the exploitation of natural resources, we start mining the talent and intelligence of human resources, necessary for the innovation that will ensure our survival. Richard Florida's work shows the centrality of cities in this process. He has mapped where innovation occurs, and it spikes dramatically in major cities, with the most inventiveness happening in those places with the greatest density of people.[14] As Ryan Avent observes in *The Gated City*, "Economist Masayuki Morikawa finds that productivity rises between 10 and 20 percent when density doubles."[15]

Nor does this apply only to the wealthy elite. As Katherine Boo recounts in her book *Behind the Beautiful Forevers* about life in the Mumbai slum, Annawadi, poor people continue to flood into cities from rural areas and live in overcrowded conditions for one reason: hope.[16] (Figure 11.4) Cities offer opportunities for people almost regardless of their financial situation or educational attainment. In the case of Annawadi, the inventiveness of the people revolves around the recycling and repurposing of the trash coming from the nearby luxury hotels and Mumbai airport. While we tend to see innovation in terms of the newest high-tech devices, the creativity of the people of Annawadi may point more to where we need to go, using everything we have in inventive new ways.

11.4 The
Annawadi
slum, Mumbai.
Photograph:
Kevin Jones

We humans, for all of our intelligence, remain one of the only animals that creates waste that other plants or animal can't use. We also stand out as the one species that doesn't draw our energy primarily from the sun, winds, or tides, but instead from the carbon buried in the earth in the form of oil and gas. The creation of mountains of trash and the shifting of carbon from the earth to the atmosphere all seems normal to us, the product of our technological prowess. But when viewed from the perspective of other species, we are very primitive indeed, using brute force to manipulate—and some would say, maul—the planet rather than using our creativity and inventiveness to utilize all of the "free" resources at our disposal. We have a lot of catching up to do with the other animals with whom we share this world.

Humanity has always thrived in dense communities. We are, as Aristotle noted a few millennia ago, among the most social of the animals, and we should stop fighting that fact with public policies that openly discourage density and perversely privilege privacy above almost all other values. The suburbanization of modern life has had all sorts of unintended consequences from increased obesity from an auto-oriented existence to high death rates among people aged 5 through 34 as a result of motor vehicle crashes to increased dependence on fossil fuels in order to meet our most basic needs.[17]

But one of the impacts that until recently has gone largely unnoticed is the negative effect that a physically isolating built environment has on our ability to innovate and create. And when innovation has occurred in dispersed settings, people have had to work against the isolation of these places. As Edward Glaeser notes in *Triumph of the City*, "The computer industry, more than any other sector, is the place where one might expect remote communication to replace person-to-person meetings … Yet despite their ability to work at long distances, this industry has become the most famous example of the benefits of geographic concentration."[18]

It matters, though, what kind of city we build. Too often, our image of the city of the future has a futuristic character that, paradoxically, works against the very innovation we need to pursue. The modern city, now rising rapidly in places like China, equates high-density development with private luxury, perpetuating the fantasy that everyone will lead high-tech lives in high-rise buildings, using high-speed transportation systems and high-energy digital networks as they move back and forth between their high-end housing and offices. It is all so 20th century.

We cannot sustain such fantasies in the 21st century if we are to avoid the distinct possibility that we may be among the 50 percent of the species headed for eradication in the "sixth extinction" that we, ourselves, have set in motion. Instead, we need to start creating, as quickly as possible, environments that maximize our ability to innovate in ways that allow us to sustain ourselves on this planet. To do so, we should keep a few principles in mind. First, we need to try to match the sophistication of almost every other species and learn to meet all of our needs within an ecological footprint of 100 watts and 2,000 calories a day. We used to live well within such limits, and many people around the world still do, so this presents a completely achievable goal—if we have the imagination enough to realize it. (Figure 11.5)

Second, we need to design everything in such as way that it utilizes renewable resources and generates no waste that either we cannot reuse or that another species cannot consume. Our ancient ancestors knew how to do this, apparent in the relative lack of archaeological evidence of their existence, and we need to reinvent our future with that past in mind.

Third, we need to reinvent ourselves and our ideas about our relationships to other species, future generations, and the planet as a whole, moving from a

11.5   Kjellgren Kaminsky Architecture in Goteborg, Sweden, are among many firms that have envisioned more environmentally friendly cities, while highlighting the unsustainable aspects of current zoning regulations and economic policies. Kjellgren Kaminsky Architecture

competitive and exploitative stance to a cooperative and mutually enhancing one. This parallels how our view of nature has changed from one focused on the survival of the fittest to one that acknowledges our interdependence and diversity within ecosystems.

THE CITY OF INNOVATION

Such principles do not preclude innovation; rather, they provide a framework within which innovation can occur. We have plenty of "innovation" going on in the world now, but much of it consists of superficial variations of what already exists or new ideas that seem to perpetuate our existing relationship with nature. And when true innovations have occurred recently, too many of them had unintended negative consequences for other people or the planet. In a framework based on exploiting resources, low-cost labor, and the habitat of other species, much of the innovation we have had is not only unsustainable, but also undermining our very ability to support ourselves over the long term. That isn't innovation; it's a kind of self-immolation.

So let's look at what it will take for us to achieve the kind of innovation that will enable us to sustain ourselves, with the same kind of sophistication that most other animals have already accomplished so brilliantly through evolution. What would an architecture and city of such innovation be like? On the surface, it might look a lot like the high-rise cities so often depicted in our futuristic fantasies. But they would function very differently than the kinds of cities we now know, at least in the most developed countries.

Unlike the economic and ethnic segregation of most cities today, a city aimed at maximizing innovation would consciously mix people of varying social and economic backgrounds. This stems not from a humanitarian interest in equity, but from a realistic assessment of human capability: intelligence, inventiveness, and creativity spreads across all socioeconomic groups and so enabling those with the best ideas to have a chance to be heard and to develop a concept that might revolutionize our ability to sustain ourselves demands that we find ways to enable the all people to have such opportunities. This cannot happen if we separate rich and poor, market-rate and subsidized housing, formal and informal settlements.

A better model—long followed in ancient cities—would mix all of it as much as possible, while obliterating any outward sign of which is which. The density of the modern high-rise city easily allows for this, with buildings that often have relatively uniform exteriors and interior public spaces. What this will require, though, is a change in public policy to encourage the integration of expensive and inexpensive housing within the same structures, perhaps through height or square-footage bonuses in exchange for this provision. Within limits, added density is a good thing in such a city.

To further maximize innovation, we would seek to minimize private space in order to maximize the public and semi-public space in which people come together to share ideas and stimulate new thinking. This, too, runs counter to the trend in most cities of increasing private space and even privatizing activities like shopping and conversing that once happened in public. While that privatization has benefited those who can afford it, it leaves out the vast majority who can't and yet who, given their overwhelming numbers, invariably have a majority of the innovative minds in their midst. The wealthy have not cornered creativity and the more we inhibit creative people, regardless of their background, the less likely we will be able to invent our way out of the dire straights in which we find ourselves.

The minimizing of private space and maximizing of public life also characterized ancient cities, which I would argue helped propel the innovations that led to modern life, from democracy and market economies to science and technology. We need to do the same again, and the most innovative cities in our own time, according to Richard Florida's assessment, already have the least amount of private space per person. Architecturally, this will demand that we look at how to make modest living quarters as functional and comfortable as possible, while questioning our assumptions about single-use rooms and single-purpose furnishings that have led to the expansion of private space. It will also require a careful look at what kinds of public spaces enhance interaction and communication. The large and often barren open spaces of the modern city, with blank street walls and unfriendly pedestrian environments, definitely do not.

Another aspect of the innovative city involves the mixing of activities in a much finer grain than what most modern cities now allow. Like the single-purpose room, single-use zoning has become one of the greatest obstacles to innovation ever invented. Separating where we live from where we work from where we study from where we shop and so on, damages our ability to learn and grow from each other, which lies at the core of our ability to create a better future for ourselves. The functionally segregated city inhibits young people from watching their elders and coming up with better ways of doing things, older people from sharing their experiences with those younger than them so that we don't make the same mistakes, and people of all ages from having the feedback loops that enable us to speed up our cycles of innovation. A city built around innovation would encourage people to live and work, study and play, produce and consume in very close proximity, ideally all within an easy walk.

Ancient cities did this well and the faster we return to the mixtures that characterized the innovative places of the past, the faster we will be able to invent the sustainable environments of the future. Architecturally, this means that buildings will need to be much more adaptable to a variety of activities rather than limited to residential, commercial, educational, or industrial uses. Such buildings will also need to have the flexibility to quickly change uses while still providing safety, security, and sound isolation. Building more flexibly on one hand, with easily recyclable elements changing as often as necessary, and building more permanently on the other, with high embodied- energy elements lasting as long as possible, would become the norm.

Such a fine-grained mix of uses has implications for the kinds of innovations we seek. We segregated activities in cities, especially those related to heavy industry, because of the danger, noise, and outright toxicity of the work going on there.

Among the things we need to innovate our way out of, ending that brute-force way of accommodating human needs must take priority. No other species rapes and pillages the planet as we have learned to do, and no other species fouls its own nest as we have done for the last few centuries. We have come to see this perverse way of being as normal and chalk it up to the price we must pay for progress, even though our ancestors spent most of human history not living like this. (Figure 11.6)

11.6 Francis Bacon, in his 17th Century book *The New Atlantis*, envisioned a society fully engaged in scientific investigation, understanding nature and developing technology in harmony with it, something we have yet to achieve some 400 years later. Illustration from The New Atlantis, Francis Bacon, 1627

Nor can we for very much longer. Pick your poison: greenhouse gases warming up the planet beyond human habitability; nuclear waste lying in wait to contaminate the air, water, and large areas of land with the next accident; or chemicals in the environment altering everything from reproductive cycles to mental development. We have never had so many ways to kill ourselves. The defenders of toxic industries make this sound alarmist and those who argue them sound paranoid, but the fact is, we do not have to live this way. We simply have not been inventive enough yet to imagine ways of meeting of all our needs without harming ourselves and the rest of planet in the process.

Our cities and buildings can lead the way here. We have long built environments for ourselves using natural materials, regional resources, and biodegradable elements, constructing our cities largely with what we had available to us locally and in sync with what the climate of a place required in terms of meeting human comfort. Only in the last 100 years or so have we overturned that successful formula, making things with materials shipped all over the world, using resources that we cannot replace once we exhaust them, and leaving a trail of waste, pollution, and environmental degradation in our wake. And only in the last 50 years or so have we created indoor environments that require enormous amounts of energy to eliminate the differences of climate or of the time of day or year. People lived happily and productively for thousands of years without depending upon such artificial and unsustainable conditions. We need to invent a future in which we do so again.

THE COUNTRY OF EXPERIMENTATION

This emphasis on the city as the source of innovation raises the question of what to do with rural areas. They, too, need to change in ways that enable us to live within our ecological footprint, and we also do not have a lot of time to make this shift. A group of scientists that included Nobel Laureate Paul Crutzen, have identified nine boundaries that, if crossed, will bring catastrophic changes to the earth's systems.[19] The nine thresholds are: climate change, stratospheric ozone, land use change, freshwater use, biological diversity, ocean acidification, nitrogen and phosphorus inputs to the biosphere and oceans, aerosol loading, and chemical pollution. The scientists believe that we have already crossed three of these boundaries— climate change, biological diversity and nitrogen input to the biosphere—and that because these boundaries have strong connections, we may have a difficult time staying within safe levels with the others.

Modern agricultural methods threaten many of these boundaries. Large-scale farming releases massive quantities of carbon dioxide into the atmosphere, propelling climate change; it alters the size and continuity of habitat, reducing biological diversity; it creates runoff that carries nitrogen and phosphorus into waterways that acidify the oceans; and it fills the air, water, and land with aerosols and chemicals, polluting the landscape. We cannot continue along this path, nor do we need to. The Food and Agriculture Organization of the United Nations has shown how we can feed 9 billion people sustainably by 2050 with more productive farming methods, more efficient agriculture technology, and better stewardship of the land and water.[20] At the same time, people's expectations need to change, moving to a largely plant-based diet and to a roughly 2,000-calorie daily intake, close to the world's average but over a third less than what (increasingly obese) North Americans eat. As in so many other areas of our lives, the transformations we need to make exist mainly in our own minds and bodies.

Rather than see rural areas as backwaters from which to escape to the city, we should view them, if we are serious about increasing the pace of innovation, as places to pilot change. While cities have long served as the caldron for innovation, their size makes it hard to bring an innovative idea up to scale in a series of steps. Rural areas and small settlements can play that role, as they have in the past, with the utopian tradition of experimental communities such as the Amana Colonies, Brook Farm, New Harmony, and the Oneida Community. (Figure 11.7) Those places largely consisted of social experiments, many of them based on religious or economic ideas. But the notion of testing new ideas at the scale of a single community before embedding them in public policy represents an important role for small settlements that would make them newly attractive for the most innovative among us.

That experimentation now needs to go beyond social relationships to include our relationship with the natural world. Relearning how to steward ecosystems

11.7   Although Robert Owen's vision for New Harmony, Indiana, failed as a communal social experiment, it did become a center for scientific discovery and a model of how rural communities can serve as the testing grounds for innovation. Photo by Marten Kuilman of 1838 illustration of New Harmony, Indiana

and bring exploited landscapes back to health constitutes one area in urgent need of exploration. Contrary to what we do now, which is incentivize farmers and rural residents to turn the land into vast commodity production facilities with monocultures of a few types of crops and animals, innovation would require that we encourage just the opposite. The more diverse the experiments and the more varied the methods, the more likely we will arrive at solutions that can sustain us at a global scale. By creating a healthier human ecosystem, with widely different patches of activity, the sooner we will create healthier natural ecosystems, with us as an integral part.

These experiments do not need to cost more money. Indeed, done well, they will show how much living in sync with nature reduces the effort needed and the economics of doing so. For example, our rural landscapes have thousands of miles of roads that connect small farmsteads and that receive very little traffic most of the time. One experiment we should start immediately would be to eliminate the least-used roads and convert their rights-of-way to more productive environmental purposes ranging from bio-filtration swales to capture and clean farm run-off to bio-fuel production containing fast-growing plants to habitat corridors that increase the biodiversity of the landscape. Such moves reduce the cost of road maintenance, generate revenue from public rights-of-way, improve environmental quality, and diversify the landscape's ecology. Rural areas are full of such opportunities, once we see the linkage between what we can imagine in cities and what we can investigate in the country.

GOING FORWARD

Much of the above will, no doubt, strike many as entirely too utopian, or as so contrary to the way we do things now that it will never come to pass except through the imposition of some politically repressive act by some wild-eyed "pinko" progressives. But it doesn't require either leftist revolution or radical politics to make this happen. We only have to wait, and not very long, for something like what I have sketched to come to pass voluntarily, as nature forces us in this direction. The relatively mild climate change we have experienced so far, with historically high temperatures and increasingly violent and unpredictable weather patterns, seems likely just the beginning of what we will face in the future. As the climatologist Mark Seeley has put it, the more energy we put into the atmosphere the more energy the atmosphere sends back to us.[21] And as a rapidly growing population continues to pursue the unsustainable path of 20th century modernity, the amount of energy we keep exhausting into the air continues to grow just as fast.

No species can so alter the ecosystems upon which It depends without paying a steep price and the bill for our ecologically destructive ways has come due. Ideally, we would have seen this coming and altered our own ways long before this, but we didn't and so we will now be forced to change. As Jared Diamond has documented in *Collapse*, communities in the past have risen to the occasion of their extinction in the face of rapid environmental change and I believe that people in at least some parts of the planet will do so again.[22] But that will demand that we let go of the old and pointlessly ideological arguments that continually hold us back—capitalism versus socialism, freedom versus regulation, class warfare versus equal opportunity, and the list goes on. Such arguments amount to fiddling while Rome—or in our case, Earth—burns.

The future lies, instead, with those who innovate their way out of this mess as fast as possible. And as I hope is apparent by now, the first innovations that need to occur involve overturning the very rules and regulations that have made innovation so difficult in the recent past. We know how to create places that encourage innovation; the history of cities offers ample examples of that. But we cannot go forward simply by looking backward. Instead, we need to imagine a future in which we catch up to the other species much older—and I think, much wiser—than humanity, learning to lead happy, fulfilling, and satisfying lives within the boundaries that the Earth can support. We can do this. And if we don't, it won't matter, since there may be none of us left to wonder why the most intelligent animal yet to walk the planet Earth could have been so stupid.

NOTES

1   United Nations, Department of Economic & Social Affairs, Population Division. "World Population to 2300." New York: United Nations, 2004.

2   Architecture 2030. "Architecture 2030 Will Change the Way You Look at Buildings," http://architecture2030.org/the_problem/buildings_problem_why

3   Leakey, Richard and Lewin Roger. *The Sixth Extinction: Patterns of Life and the Future of Mankind.* New York: Anchor Books, 1996.

4   Design Other 90 Network. http://www.designother90.org/

5   Global Footprint Network. http://www.footprintnetwork.org/

6   UN Habitat. *State of the World's Cities, 2010–2011, Bridging the Urban Divide.* New York, United Nations, 2011.

7   "Pandemic (H1N1) 2009," *Global Alert and Response.* World Health Organization. www.who.int/csr/disease/swineflu/en/index.html

8   U.S. Environmental Protection Agency. "Climate Change Impacts and Adpating to Change." http://www.epa.gov/climatechange/impacts-adaptation/

9   Fisher, Thomas. *Designing to Avoid Disaster, The Nature of Fracture-Critical Design.* London: Routledge, 2012.

10   Klein, Naomi. *The Shock Doctrine: The Rise of Disaster Capitalism.* New York: Henry Holt, 2007.

11   West, Geoffrey. "The Universal Scale of Life." *Thought Leader Forum,* 2002. www.capatcolumbia.com/CSFB%20TLF/2002/west_sidecolumn.pdf

12   West, Geoffrey. "Why the Future of Humanity and the Long-term Sustainability of the Planet are Inextricably Linked to the Fate of our Cities." *Seed.* July 5, 2010

13   Friedman, Thomas. *New York Times.* "Pass the Books. Hold the Oil." New York: March 10, 2012.

14   Florida, Richard. *Who's Your City, How the Creative Economy is Making Where You Live the Most Important Decision of your Life.* New York: Basic Books, 2008.

15   Avent, Richard. *The Gated City, How America Made its Most Productive Place Ever Less Accessible.* Lulu Press, 2011.

16   Boo, Katherine. *Behind the Beautiful Forevers: Life, Death, and Hope in a Mumbai Undercity.* New York: Random House, 2012.

17   Centers for Disease Control and Prevention. "Injury Prevention & Control: Motor Vehicle Safety." http://www.cdc.gov/motorvehiclesafety/

18   Glaeser, Edward. *Triumph of the City: How our Greatest Invention Makes Us Richer, Smarter, Greener, Healthier, and Happier.* New York: Penguin, 2011.

19   Crutzen, Paul et al. *Ecology and Society. "Planetary Boundaries: Exploring the Safe Operating Space for Humanity."* Vol. 14, #2, 2009. http://www.ecologyandsociety.org/vol14/iss2/art32/

20   United Nations Food and Agriculture Organization. "How to Feed the World in 2050." http://www.fao.org/wsfs/forum2050/wsfs-forum/en/

21   Seeley, Mark. From a talk given as the Association of Collegiate Schools of Architecture, Minneapolis, MN. 2007.

22   Diamond, Jared. *Collapse, How Societies Choose to Fail or Succeed.* New York: Viking, 2005.

Bibliography

Adams, Frederick J, *Manual of Office Practice for the Architectural Worker*, (New York: Charles Scribner's Sons, 1924).

Allen, Stan and Marc McQuade, *Landform Building: Architecture's New Terrain*, (Baden: Lars Muller; New Jersey: Princeton University School of Architecture, 2011).

American Institute of Architects, "The Business of Architecture: 2003 AIA Firm Survey," American Institute of Architects in partnership with McGraw-Hill Construction (2003).

American Institute of Architects, *The Handbook of Architectural Practice,* (Washington, D.C.: AIA, 1943).

American Institute of Architects, *The Handbook of Architectural Practice,* (Washington, D.C.: AIA, 1958).

American Institute of Architects, *The Handbook of Architectural Practice*, (Washington, D.C.: AIA, 1927).

Anderson, Mark, *Prefab Prototypes: Site Specific Design for Off-Site Construction*, (Princeton Architectural Press, 2007).

Annual Report on Relocatable Buildings, Modular Building Institute, 2011.

Annual Report on Permanent Modular Construction, Modular Building Institute, 2011.

"Apple's Move Keeps Profits out of Reach of Taxes," *New York Times*, (2 May 2013).

"Apple's Web of Tax Shelters Saved It Billions, Panel Finds," *New York Times*, (20 May 2013).

Appy, Christian G., (ed.), *Cold War Constructions: The Political Culture of United States Imperialism, 1945–1966* (Amherst: University of Massachusetts, 2000).

"The Architect as Developer," *Architectural Record*, 10, (Oct 1971): 126–127.

Architecture for Humanity, Open Architecture Challenge: Classroom, 2009.

Ascher, Charles A., "Survival through Design by Richard Neutra," *Academy of Political and Social Sciences* (September, 1954): 182–183.

Avent, Richard, *The Gated City, How America Made its Most Productive Place Ever Less Accessible*, (Lulu Press, 2011).

Awan, Nishat, Tatjana Schneider, and Jeremy Till, (eds), *Spatial Agency: Other Ways of Doing Architecture*, (Routledge, 2011).

BAVO, eds., *Cultural Activism Today: The Art of Over-Identification*, (Episode Publishers, 2007).

"Behind the Rise in House Prices, Wall Street Buyers," *New York Times*, (4 June 2013).

Bell, Bryan and Wakeford, Katie, *Expanding Architecture: Design as Activism*, (Metropolis Books, 2008).

Benjamin, Water, "The Author as Producer," in *Reflections: Essays, Aphorisms, Autobiographical Writings*, (Schocken, 1986).

Benjamin, Walter, "Paris, Capital of the Nineteenth Century," in *Reflections: Essays, Aphorisms, Autobiographical Writings*, ed. P. Demetz (Schocken, 1986).

Berman, Marshall, *All that is Solid Melts into Air*, (Penguin, 1988).

Blake, Peter, *No Place Like Utopia: Modern Architecture and the Company We Kept*, (New York: Knopf, 1993).

Bledstein, Burton J., *The Culture of Professionalism: The Middle Class and the Development of Higher Education in America*, (New York: Norton, 1976).

Boo, Katherin, *Behind the Beautiful Forevers: Life, Death, and Hope in a Mumbai Undercity*, (New York: Random House, 2012).

Borges, Jorge Luis, *The Book of Imaginary Beings*, (USA: Penguin Group, 2006).

Chafe, William H. and Harvard Sitkoff (eds), *A History of Our Time: Readings on Postwar America*, (5th edn., New York 1999).

Childs, Marquis W., *Mighty Mississippi: Biography of a River*, (New Haven: Ticknor & Fields, 1982).

Civil Defense: The Architect's Part, (Washington, DC: AIA, 1951).

Coates, Ta-Nehsi, "The Littlest Schoolhouse," *Atlantic Monthly*, (Jul/Aug 2010).

Colvin, Howard, *A Biographical Dictionary of British Architects 1660–1840*, (London: J. Murray, 1954).

Conard, Mark T., *The Philosophy of Martin Scorsese,* (Lexington: University Press of Kentucky, 2007).

"The Corrosive Effect of Apple's Tax Avoidance," *New York Times*, (23 May 2013).

Crutzen, Paul et al. *Ecology and Society*. "*Planetary Boundaries: Exploring the Safe Operating Space for Humanity,*" 14(2), (2009).

Cuff, Dana and John Wriedt (eds), "Architecture: The Entrepreneurial Profession," in *Architecture from the Outside In: Selected essays by Robert Gutman*, (New York: Princeton Architectural Press, 2010).

Davis, Mike, *City of Quartz: Excavating the Future in Los Angeles*, (New York: Vintage, 1992).

De Hart, Jane Sherron, "Containment at Home: Gender, Sexuality, and National Identity in Cold War America," in *Rethinking Cold War Culture*, Peter J. Kuznick and James Gilbert (eds), (Washington: Smithsonian Institution, 2001).

DeLorenzo, Ike, "A First Course Gets High Grades," *The Boston Globe*, (29 December 2010).

Demkin, Joseph A. (ed.), *Security Planning and Design: A Guide for Architects and Building Design Professionals*, (Hoboken, N.J.: J. Wiley & Sons, 2004).

Dewey, John, "My Pedagogic Creed," in *School Journal*, 54, (January 1897).

Diamond, Jared. *Collapse, How Societies Choose to Fail or Succeed*, (New York: Viking, 2005).

"'Dream a Little Dream,' at $250,000 a Minute," *New York Times*, (23 June 2013).

Dudley, Michael Quinn, "Sprawl as Strategy: City Planners Face the Bomb," *Journal of Planning Education and Research*, 21(1), (September 2001): 52–63.

Duliere, Aude-Line and Clara Wong, *Monsterpieces*, (USA: ORO editions, 2010).

Eco, Umberto, *The Infinity of Lists*, (New York: Rizzoli, 2009).

Ellin, Nan (ed.), *The Architecture of Fear*, (New York: Princeton Architectural Press, 1997).

"Excerpts from Survival through Design, by Richard Neutra," *Architectural Forum*, 100, (Jan 1954): 130–133.

"Facebook Paid £2.9m Tax on £840m in Profits Made Outside the US, Figures Show," *The Guardian*, (23 December 2012).

Farish, Matthew, "Disaster and Decentralization: American Cities and the Cold War," *Cultural Geographies*, 10/2, (2003): 125–148.

Fickes, Michael, "8 Ways to Protect Your Building during a Demonstration," accessed May 18, 2012, http://www.buildings.com/tabid/3334/ArticleID/13382/Default.aspx.

Fickes, Michael, "Guess Who's Coming into Your Building," accessed May 18, 2012, http://www.buildings.com/tabid/3334/ArticleID/11089/Default.aspx.

Fickes, Michael, "Protect Buildings from Vehicle Attacks," accessed May 18, 2–12, http://www.buildings.com/tabid/3334/ArticleID/13693/Default.aspx.

Fisher, Thomas. *Designing to Avoid Disaster, The Nature of Fracture-Critical Design*, (London: Routledge, 2012).

Fisher, Thomas, "The Once and Future Profession," *Archvoices, AIA Online*, (2002).

Florida, Richard. *Who's Your City, How the Creative Economy is Making Where You Live the Most Important Decision of your Life*, (New York: Basic Books, 2008).

Forty, Adrian, "Form" in *Words and Buildings, A Vocabulary of Modern Architecture*, (Thames & Hudson, 2000).

Forty, Adrian, *Objects of Desire*, (Thames and Hudson, 1986).

Friedman, Avi, "Developing Skills for Architects of Speculative Housing," *Journal of Architectural Education*, 47(1), (Sep 1993): 49–52.

Friedman, Thomas , "Kicking Over the Chessboard," *New York Times*, (18 April 2004).

Friedman, Thomas, "Pass the Books. Hold the Oil," *New York Times*, (10 March 2012).

Fulcher, Merlin, "Prefab Schools Debate Heats Up," *Architects Journal,* October (2010).

Gibbs, Nancy, "The Aftermath," *Time: An American Tragedy*, (12 September 2005).

Gissen, David, *Subnature: Architecture's Other Environments*, (New York: Princeton Architectural Press, 2009).

Glaeser, Edward. *Triumph of the City: How our Greatest Invention Makes Us Richer, Smarter, Greener, Healthier, and Happier*, (New York: Penguin, 2011).

Grafley, Dorothy, "Survival through Design," Weathervane section, *American Artist*, (November 1954): 48, 56–58.

Graham, Steven (ed.), *Cities, War, and Terrorism: Towards an Urban Geopolitics*, (Oxford: Blackwell, 2004).

Gray, Mitchell and Elvin Wyly, "The Terror City Hypothesis," in *Violent Geographies: Fear, Terror, and Political Violence*, Derek Gregory and Allan Pred (eds), (New York: Routledge, 2007).

Hackworth, Jason, *The Neoliberal City*, (Ithaca, NY: Cornell University Press, 2006).

Hanczyc, M.M., T. Toyota, T. Ikegami, N. Packard and T. Sugawara, "Fatty Acid Chemistry at the Oil-Water Interface: Self-Propelled Oil Droplets," *Journal of The American Chemical Society,* 129(30): (2007): 9386–91.

Harvey, David, *Paris, Capital of Modernity,* (Routledge, 2005).

Hawthorne, Christopher, "Prefab: the Dream that Refused to Die," *Metropolis,* (2011).

Hay, James, "Designing Homes to be the First Line of Defense," *Cultural Studies,* 20(4), (2006).

Heck, J.G., *The Complete Encyclopedia of Illustration,* (New York: Park Lane, 1979).

Heerwagen, Judith, "Psychosocial Value of Space," *Whole Building Design Guide,* (National Institute of Building Science, 2008).

Henry, Edward William, *Portman, Architect and Entrepreneur. The Opportunities, Advantages, and Disadvantages of His Design Development Process,* (Ann Arbor, MI: University Microfilms, 1985).

Hines, Thomas, *Richard Neutra and the Search for Modern Architecture,* (Rizzoli: New York, 2005).

Hirst, Paul, *Space and Power: Politics, War and Architecture,* (Cambridge, UK: Polity Press, 2005).

Hopper, Leonard J. and Martha J. Droge, *Security and Site Design: A Landscape Architectural Approach to Analysis, Assessment, and Design Implementation,* (Hoboken, N.J.: J. Wiley & Sons, 2005).

Imrie, Rob. 2004. "Disability, Embodiment and the Meaning of the Home," *Housing Studies,* 19(5).

Ivan, Johnson, "Survival through Design by Richard Neutra," *Marriage and Family Living,* 17, (1955): 181–182.

Jencks, Charles, "Hetero-Architecture for the Heteropolis: The Los Angeles School," in Nan Ellin (ed.), *The Architecture of Fear,* (New York: Princeton Architectural Press, 1997).

Johnson, Kirk, "School District Bets Future on Real Estate," *New York Times,* (4 Sept. 2012).

Johnston, George Barnett, "Professional Practice: Can Professionalism be Taught in School," in Joan Ockman (ed.), *Architecture School: Three Centuries of Educating Architects in North America,* (Cambridge: MIT Press, 2012).

"Justice Still Elusive in Factory Disasters in Bangladesh," *New York Times,* (29 June 2013).

Katz, Cindi, "Banal Terrorism: Spatial Fetishism and Everyday Insecurity," in Derek Gregory and Allan Pred (ed.), *Violent Geographies: Fear, Terror, and Political Violence,* (New York: Routledge, 2006).

Kemper, James Parkerson, *Rebellious River,* (Boston: Humphries, 1949).

Klein, Naomi. *The Shock Doctrine: The Rise of Disaster Capitalism,* (New York: Henry Holt, 2007).

Kline, Morris. "The Sine of G Major," *Mathematics in Western Culture,* (Oxford: Oxford University Press, 1953).

Koolhaas, Rem, "Bigness: The Problem of Large," in *Wiederhall,* 17, (1994).

Kostof, Spiro, (ed.), *The Architect: Chapters in the History of the Profession,* (New York: Oxford University Press, 1977).

Krauss, Rosalind E., "Clock Time," *October,* 136, (Spring 2011).

Lavin, Sylvia, *Form Follows Libido, Architecture and Richard Neutra in a Psychoanalytic Culture*, (MIT Press, 2005).

Leakey, Richard and Lewin Roger. *The Sixth Extinction: Patterns of Life and the Future of Mankind*, (New York: Anchor Books, 1996).

Leatherbarrow, David, *Uncommon Ground, Architecture, Technology and Topography*, (MIT Press, 2000).

Linder, Mark, "Disciplinarity: Redefining Architecture's Limits and Identity," in Joan Ockman (ed.), *Architecture School: Three Centuries of Educating Architects in North America*, (Cambridge: MIT Press, 2012).

Low, Setha, "The Fortification of Residential Neighborhoods and the New Emotions of Home," Special Issue, M. Van de Land and L. Reinders (eds) *Housing, Theory and Society*, 25(1): 47–65.

"Luxe Builders Chase Dreams of Tax Breaks," *New York Times*, (25 June 2013).

Machado, Rodolfo, *Monolithic Architecture,* (Munich: Prestel, 1995).

Mallory, Keith and Arvid Ottar, *Architecture of Aggression: A History of Military Architecture in North West Europe 1900–1945*, (London, UK: Architectural Press 1973).

Marcuse, Peter, "Urban form and globalization after September 11th: the view from New York," *International Journal of Urban and Regional Research*, 26(3), (September 2002).

Marx, Karl, *Capital, Volume One*, (Penguin, 1992 [1867]).

Marx, Karl, "Economic and Philosophic Manuscripts of 1844," in R. Tucker (ed.), *Marx-Engels Reader*, (Norton, 1978).

Menges, A. 2011. Integrative Design Computation: Integrating material behavior and robotic manufacturing processes in computational design for performative wood constructions. *ACADIA 2011: Integration through Computation*, (Banff, Alberta): 72–81.

Miles, Lawrence D., *Techniques of Value Analysis and Engineering*, (New York: McGraw-Hill Book Company Inc., 1961).

Monteyne, David, *Fallout Shelter: Designing for Civil Defense in the Cold War*, (Minneapolis: University of Minnesota Press, 2011).

Morton, Jennie, "Corporate Security Plan Tips," accessed May 18, 2012, http://www.buildings.com/tabid/3334/ArticleID/12966/Default.aspx.

Moser, Cliff, "Using Active Value Engineering for Quality Management," The American Institute of Architects, (2009).

Mostafavi, Mohsen and Gareth Doherty, (eds), *Ecological Urbanism*, Harvard University, Graduate School of Design, (Baden: Lars Müller Publishers, 2010).

Nadel, Barbara (ed.), *Building Security: Handbook for Architectural Planning and Design*, (New York: McGraw-Hill, 2004).

Nagy, Sibyl Moholy, "Environment and Anonymous Architecture," *Perspecta*, 3, (1955): 2–7, 77.

Nagy, Sibyl Moholy, "Survival through Design by Richard Neutra," *College Art Journal*, 13(4), (Summer 1954): 329–331.

Nasr, Joe "Planning Histories, Urban Futures, and the World Trade Center Attack," *Journal of Planning History*, 2(3), (August 2003): 195–211.

Neutra, Richard, *Survival through Design*, (Oxford University Press, 1954–1969).

Nichols, Frederick Doveton, "Jefferson: The Making of an Architect," in William Howard Adams (ed.), *Jefferson and the Arts: An Extended View*, (Washington: National Gallery of Art, 1976).

Norris, Floyd, "Masked by Gibberish, Risks Run Amok," *New York Times*, (21 March 2013).

Palladio, Andrea, *The Four Books of Architecture*, (New York: Dover Publications, Inc., 1965).

Pallasmaa, Juhani, *The Embodied Image: Imagination and Imagery in Architecture*, (United Kingdom: John Wiley & Sons Ltd., 2011).

Pallasmaa, Juhani, "Toward an Architecture of Humility: On the Value of Experience," in W. Saunders (ed.), *Judging Architectural Value*, (University of Minnesota Press, 2007).

Pangle, Thomas L., "The Philosophic Understandings of Human Nature Informing the Constitution," in Allan Bloom (ed.), *Confronting the Constitution: The Challenge to Locke, Montesquieu, Jefferson, and the Federalists from Utilitarianism, Historicism, Marxism, Freudianism, Pragmatism, Existentialism.*, (Washington, D.C.: The AEI Press, 1990).

Pearson, Jason, (ed.), *University-Community Design Partnerships: Innovations in Practice*, (National Endowment for the Arts, 2002).

Perez-Gomez, Alberto, *Architecture and the Crisis of Modern Science*, (Cambridge, Massachusetts: The MIT Press, 1983).

Perez-Gomez, Alberto, Introduction, *Ordonnance for the Five Kinds of Columns after the Method of the Ancients/Claude Perrault* (Santa Monica, CA: Getty Center for the History of Art and the Humanities, 1993).

Perkins, Constance M., "Survival through Design by Richard Neutra," *Journal of Aesthetics and Art Criticism*, 13, (1954): 273–74.

Place, Charles A., *Charles Bulfinch: Architect and Citizen*, (New York: Da Capo Press, 1968).

Pollio, Marcus Vitruvius, *De Architectura*, trans. Frank Granger, (Cambridge, Massachusetts: Harvard University Press, 1983).

Pollio, Marcus Vitruvius, *De Architectura*, trans. Hicky Morgan, (Dover, 1960).

Pollio, Marcus Vitruvius, *De Architectura*, trans. Ingrid D. Rowland and Thomas Noble Howe, (Cambridge University Press, 1999).

Pollio, Marcus Vitruvius, *De Architectura*, trans. Richard V Schofield's translation, (Penguin, 2009).

Porter, Michael, *Competitive Advantage: Creating and Sustaining Superior Performance*, (New York: Free Press, 1985).

Porter, Michael, *Competitive Strategy*, (New York: Free Press, 1980).

Portman, Jonathan and Jonathan Barnett, *The Architect as Developer*, (New York: McGraw-Hill, 1976).

Pfretzschner, Paul A. "Survival through Design by Richard Neutra," *The Western Political Quarterly*, 8, (March 1955): 146–147.

Riether, Gernot, Knox, J. 2011. Flexible Systems: Flexible Design, Material and Fabrication: The AIA pavilion as a case study. *eCAADe 2011: Respecting Fragile Places*, (Ljubljana, Slovenia): 628–634.

Riether, Gernot, 2012. Pavilion for New Orleans. *DETAIL /2012* ('Vorfertigung'): 6l, 613, 642–644.

Riether, Gernot, 2011. Adaptation: A pavilion for the AIA in New Orleans. *ACADIA 2011: Integration through Computation*, (Projects), (Banff, Alberta): 52–57.

RIBAJ Editors, "Architect/Developer John Portman," *RIBA Journal*, 12, (December 1977): 504–509.

Richardson, Nathan, "Architecture & Enterprise: A History, Practice, and Analysis of Architectural Extensions into Real Estate," (master's thesis, Harvard University Graduate School of Design, 2009).

Rybczynski, Witold, *Home: A Short History of an Idea*, (Viking, 1986).

Saint, Andrew, *Architect and Engineer: A Study in Sibling Rivalry*, (New Haven: Yale University Press, 2007).

Saint, Andrew, *The Image of the Architect*, (New Haven: Yale University Press, 1983)

Sheine, Judith, *R.M.Schindler*, (Phaidon, 2001).

Sheldon, Garrett Ward, *The Political Philosophy of Thomas Jefferson*, (Baltimore: Johns Hopkins University Press, 1991).

"Sky High and Going Up Fast: Luxury Towers Take New York," *New York Times*, (19 May 2013)

Smith, Neil, *The New Urban Frontier: Gentrification and the Revanchist City*, (New York: Routledge, 1996).

Sokolove, Michael, "Busted: Can the Model for Casino Gambling be Fixed?" in *The New York Times Magazine*, (18 March 2012).

Sorkin, Michael (ed.), *Variations on a Theme Park: The New American City and the End of Public Space*, (New York: Hill & Wang, 1992).

Stephens, Suzanne and Clifford Pearson, "Millennium Part Two: Futures to Come" in *Architectural Record*, (December 1999).

Stevenson, H.H. and David E. Gumpert. "The heart of entrepreneurship," *Harvard Business Review*, 63(2), (March 1985): 85–94.

Stevenson, H.H. and J.C. Jarillo, "A Paradigm for Entrepreneurship: Entrepreneurial Management," *Strategic Management Journal*, 11, (1990): 17–27.

Summerson, John, *The Life and Work of John Nash, Architect*, (Cambridge: MIT Press, 1980).

"Survival through Design," *USA Tomorrow*, 1, (Oct 1954): [74]-79.

Taatila, Vesa P., "Learning Entrepreneurship in Higher Education," *Education + Training*, 52(1), 48–61.

Taylor, William C., *Practically Radical*, (New York: HarperCollins, 2011).

"The Trouble with Taxing Corporations," *New York Times*, (28 May 2013).

Trubiano, Franca, "High Performance Homes: Metrics, Ethics and Design," in *Design and Construction of High Performance Homes – Building Envelopes, Renewable Energies and Integrated Practice*, (Routledge Press, 2012).

Twain, Mark, *Adventures of Huckleberry Finn*, (1885. New York: Dover Publications, 1994).

Twain, Mark, *Life on the Mississippi River* (1883. New York: Random House, 2007).

Tyler May, Elaine, *Homeward Bound: American Families in the Cold War Era* (Rev. edn.), New York: Basic, 1999).

United Nations, Department of Economic & Social Affairs, Population Division, "World Population to 2300," (New York: United Nations, 2004).

UN Habitat. *State of the World's Cities, 2010–2011, Bridging the Urban Divide.* (New York, United Nations, 2011).

Vale, Lawrence J., "Securing Public Space," *Places* 17(3), (2005): 38–42.

Vidler, Anthony, "A City Transformed: Designing 'Defensible Space,'" *Grey Room*, 7, (Spring 2002): 82–85.

von Uexküll, Jakob, "A Stroll Through the Worlds of Animals and Men: A Picture Book of Invisible Worlds," *Instinctive Behavior: The Development of a Modern Concept*, ed. and trans. Claire H. Schiller, (New York: International Universities Press, Inc., 1957).

Wagener, Wolfgang, *Raphael Soriano*, (Phaidon, 2002).

Waldheim, Charles, *The Landscape Urbanism Reader*, (New York: Princeton Architectural Press, 2006).

Ward, Anthony, "The Suppression of the Social in Design," in Dutton and Mann (ed.), *Reconstructing Architecture: Critical Discourses and Social Practices,* (University of Minnesota Press, 1996).

Weeks, Katie, "Designers of the Year, John Peterson and John Cary, Public Architecture," *Contract Magazine*, 50(1), (January 2009).

Weinstock, M., "Morphogenesis and the Mathematics of Emergence," in Jencks, C., and Kropf, K., (eds), *Theories and Manifestos, of Contemporary Architecture,* (2006, Chichester, West Sussex, Wiley).

West, Geoffrey. "Why the Future of Humanity and the Long-term Sustainability of the Planet are Inextricably Linked to the Fate of our Cities," *Seed,* July 5, 2010.

Wilkerson, Isabel, "The River Untamable," *The New York Times*, (08 May 2011).

Wilson, E.O., *Biophilia,* (Harvard University Press, 1984).

Zalewski, Daniel, "The Hours: How Christian Marclay created the ultimate digital mosaic," *The New Yorker*, (12 March 2012).

Index

T - #0527 - 071024 - C192 - 246/174/9 - PB - 9780367739621 - Gloss Lamination